Lecture Notes in Mathematics 1539

Editors:
A. Dold, Heidelberg
B. Eckmann, Zürich
F. Takens, Groningen

Subseries: Institut de Mathématiques, Université de Strasbourg

Adviser: P. A. Meyer

T0222669

Michel Coornaert   Athanase Papadopoulos

Symbolic Dynamcis
and Hyperbolic Groups

Springer-Verlag Berlin Heidelberg GmbH

Authors

Michel Coornaert
Athanase Papadopoulos
Institut de Recherche Mathématique Avancée
Université Louis Pasteur et CNRS
7, rue René Descartes
F-67084 Strasbourg, France

Mathematics Subject Classification (1991): 53C23, 34C35, 54H20, 58F03, 20F30

ISBN 978-3-540-56499-7   ISBN 978-3-540-47573-6 (eBook)
DOI 10.1007/978-3-540-47573-6

Typesetting: Camera ready by author
46/3140-543210 - Printed on acid-free paper

A Martine,

à Marie Pascale.

Table of contents

Introduction   1

Chapter 1. — A quick review of Gromov hyperbolic spaces.   5

§1. — Hyperbolic metric spaces.   6
§2. — Hyperbolic groups.   8
§3. — The boundary of a hyperbolic space.   9
§4. — The visual metric on the boundary.   12
§5. — Approximation by trees.   13
§6. — Quasi-geodesics and quasi-isometries.   13
§7. — Classification of isometries.   16
§8. — The polyhedron $P_d(X)$.   17
Bibliography for Chapter 1.   18

Chapter 2. — Symbolic dynamics.   19

§1. — Bernoulli shifts.   20
§2. — Expansive systems.   22
§3. — Subshifts of finite type.   26
§4. — Systems of finite type and finitely presented systems.   29
§5. — Symbolic dynamics on $\mathbb{N}$ and on $\mathbb{Z}$.   31
§6. — Sofic systems.   36
Notes and comments on Chapter 2.   40
Bibliography for Chapter 2.   41

Chapter 3. — The boundary of a hyperbolic space as a finitely presented dynamical system.   43

§1. — The cocycles φ.   44
§2. — Integration of the cocycles φ.   45
§3. — The cocycles associated to Busemann functions.   46
§4. — The gradient lines defined by φ.   50
§5. — The point at infinity associated to a cocycle.   52
§6. — Properties of the map $a : \Phi \to \partial X$.   55
§7. — Integral cocycles on a hyperbolic graph.   59

§8. — A finite presentation of the boundary of a hyperbolic group.   62
Notes and comments on Chapter 3.   67
Bibliography for Chapter 3.   68

Chapter 4. — Another finite presentation for the action of a hyperbolic group on its boundary.   69
§1. — Convergent sequences.   70
§2. — Convergent quasi-geodesic fields.   74
§3. — Integral fields on a graph.   83
§4. — Another finite presentation of the boundary of a hyperbolic group.   85
Notes and comments on Chapter 4.   89
Bibliography for Chapter 4.   90

Chapter 5. — Trees and hyperbolic boundary.   91
§1. — Trees and projective sequences of sets.   92
§2. — The tree $T_{geo}(X)$.   93
§3. — The tree $T_{part}(X)$.   96
§4. — The tree T_{lex} associated to a hyperbolic group.   100
§5. — A remark on the dimension of the hyperbolic boundary.   103
Notes and comments on Chapter 5.   105
Bibliography for Chapter 5.   106

Chapter 6. — Semi-Markovian spaces.   107
§1. — Semi-Markovian spaces.   108
§2. — First examples of semi-Markovian spaces.   11
§3. — Semi-Markovian presentation of polyhedra.   11
Notes and comments on Chapter 6.   11
Bibliography for Chapter 6.   11

Chapter 7. — The boundary of a torsion-free hyperbolic group as a semi-Markovian space.   11
§1. — N-equivalence in the trees T_{geo} and T_{part}.   11
§2. — The boundary of T_{geo} as a semi-Markovian subset.   1
§3. — A semi-Markovian presentation of $\partial\Gamma$.   1
§4. — The boundary of T_{part} as a semi-Markovian subset.   1
§5. — A finite-to-one semi-Markovian presentation of $\partial\Gamma$.   1
Notes and comments on Chapter 7.   1
Bibliography for Chapter 7.   1

Index.   1

Introduction

Gromov's theory of hyperbolic groups studies finitely generated groups whose structure has the flavour of a certain "geometry of negative curvature", when these groups are equipped with the word metric relative to an arbitrary finite generating set. In these notes, we develop a few aspects of this theory, which are related to the theory of dynamical systems. The theory of dynamical systems is understood here in a topological sense, more precisely from the point of view of "symbolic dynamics".

Symbolic dynamics studies the action of a group (or, more generally, of a semigroup) on the set of maps from this group into a finite set, which is called the set of symbols. Such an action is called a "Bernoulli shift". One can consider these actions as "building blocks" for more general actions (subshifts and their quotients), and symbolic dynamics can be considered more generally as the study of the action of a group on a space by comparing this action to one of these Bernoulli shifts.

The central object in this study will be the boundary $\partial\Gamma$ of a hyperbolic group Γ. This boundary is a compact metrizable space equipped with a continuous action of Γ. The dynamical system $(\partial\Gamma, \Gamma)$ is canonically associated to the abstract group Γ. The following questions arise naturally in this context:

1) What can we say about the topology of $\partial\Gamma$?

2) What can we say about the dynamical system $(\partial\Gamma, \Gamma)$?

Let us begin by noting that up to conjugacy, there is only a countable set of such systems $(\partial\Gamma, \Gamma)$. Indeed, every hyperbolic group is finitely presented, and the class of isomorphism classes of hyperbolic groups is therefore countable.

The behaviour of a Kleinian group on the sphere at infinity of hyperbolic space — behaviour whose study goes back to Poincaré — suggests that the action of a hyperbolic group on its boundary is, in general, "chaotic". The example of free groups of finite rank ≥ 2, for which the boundary is a Cantor set, shows that the topology of the boundary of a hyperbolic group can be in itself very chaotic (see also the examples, described in the thesis of Nadia Benakli, of hyperbolic groups whose boundary is a Menger curve or a Sierpinski curve). The results of Gromov which we present here reveal a "symbolic order" which organizes and structures this chaos of the boundary.

Before stating the results, let us give a few definitions from symbolic dynamics.

Consider the group $\mathbb{Z}$ of integers, and let S be a finite set of "symbols". We define $\Sigma = \Sigma(\mathbb{Z}, S)$ as the set of all maps from $\mathbb{Z}$ to S. In other words, Σ is the set of sequences of the form $\sigma = (x_i)_{i \in \mathbb{Z}}$ where $x_i \in S$. The sets $\mathbb{Z}$ and S are equipped with their discrete topology, and Σ of the product topology. The space Σ is a Cantor set if $card\ S \geq 2$.

Let us define an action of the group $\mathbb{Z}$ on Σ by the following formula:

$$\gamma(x_i) = (x_{\gamma+i})$$

for every $(x_i) \in \Sigma$ and $\gamma \in \mathbb{Z}$. The space Σ, equipped with this action of $\mathbb{Z}$, is called the *two-sided Bernoulli shift* with set of symbols S.

We say that a subset $\Phi \subset \Sigma$ is a *subshift of finite type* if there exists an integer l and a set W of words of length l on the alphabet S such that Φ is the set of sequences $(x_i)_{i \in \mathbb{Z}} \in \Sigma$ satisfying the following condition:

$\forall k \in \mathbb{Z}$, the subsequence $x_k, x_{k+1}..., x_{k+l-1}$ is in W.

Thus, a subshift of finite type is defined by "local" conditions. Let us note that a subshift of finite type is closed in Σ and is invariant by the action of $\mathbb{Z}$.

More generally, if Γ is a countable semigroup, we define the Bernoulli shift $\Sigma(\Gamma, S)$ as being the set of all maps from Γ to the set S, and we equip as before $\Sigma(\Gamma, S)$ with the product topology. In the same manner as in the particular case where $\Gamma = \mathbb{Z}$, there is a natural left action of Γ on $\Sigma(\Gamma, S)$, and we can define the associated subshifts of finite type.

An arbitrary dynamical system (Ω, Γ) (with Ω a compact metrizable space and Γ a countable semigroup acting on Ω) is said to be *of finite type* if we can find a finite set S, a subshift of finite type $\Phi \subset \Sigma(\Gamma, S)$ and a map $\pi : \Phi \to \Omega$, which is surjective, continuous and Γ-equivariant.

Finally, we shall say that the system (Ω, Γ) is finitely presented if there exists a map π defined as above, and if moreover the "kernel" of π, that is to say, the set

$$R(\pi) = \{(\sigma_1, \sigma_2) \in \Phi \times \Phi \text{ with } \pi(\sigma_1) = \pi(\sigma_2)\} \subset \Sigma \times \Sigma,$$

2

is a subshift of finite type (the product $\Sigma \times \Sigma$ being identified in a natural way with the Bernoulli shift $\Sigma(\Gamma, S \times S)$). Such a map π is called a *finite presentation* of the dynamical system (Ω, Γ).

We can make here an analogy with combinatorial group theory. Bernoulli shifts correspond to free groups of finite rank, dynamical systems of finite type to finitely generated groups, and finally finitely presented dynamical systems to finitely presented groups. The conditions of being "of finite type" or "finitely presented" is expressed, in the theory of groups as well as in that of dynamical systems, by a finite number of combinatorial rules, and this is one of the main reasons for which the study of these groups, as well as that of finitely presented dynamical systems admit a combinatorial geometric approach to which we have no access in the most general case.

Smale's Axiom A on basic sets, Thurston's peudo-Anosov maps on surfaces and Weiss's sofic systems give examples of finitely presented systems with $\Gamma = \mathbb{Z}$.

We shall see that for every hyperbolic group Γ, the dynamical system $(\partial \Gamma, \Gamma)$ is finitely presented. As in the theory of groups, it is interesting to describe several finite presentations of the same finitely presented dynamical system. We shall give two different presentations of the dynamical system $(\partial \Gamma, \Gamma)$.

Following ideas of Gromov, we develop then a "combinatorial" theory of compact spaces. This is the theory of "semi-Markovian spaces". One can consider this theory as being part of a more general theory of "symbolic topology". Using the same notations as above, we shall say that a subset of $\Sigma(\Gamma, S)$ is *semi-Markovian* if it is the intersection of a subshift of finite type with a cylinder of $\Sigma(\Gamma, S)$. A compact space Ω is said to be *semi-Markovian* if there exists a finite set S, a semi-Markovian subset $\Phi \subset \Sigma = \Sigma(\mathbb{N}, S)$ and a continuous and surjective map $\pi : \Phi \to \Omega$ whose kernel $R(\pi)$ (as defined above) is a semi-Markovian subset of $\Sigma \times \Sigma$. The definition of a semi-Markovian space does not involve any group action on the given space, and therefore, no equivariance condition is required. We shall see in particular that any finite polyhedron is semi-Markovian. Another example of a semi-markovian space which is of particular interest for us is the boundary of a torsion-free hyperbolic group . The proof uses the notion of "N-type" in a group, which has been introduced by J. Cannon (before Gromov's theory of hyperbolic groups) and which we shall recall.

Let us remark by the way that there is a different approach to the study of hyperbolic groups in the framework of dynamical systems; this is the approach that uses the probabilistic point of view, and which involve some measure theory on the boundary of a hyperbolic group (or more generally, of a hyperbolic space) equipped with the action of the group (resp. with the action of a discrete group of isometries of the space). See for example the works of M. Coornaert and of V. Kaimanovitch. This probabilistic approach is left aside in this study.

This text can be considered as a sequel to the notes on hyperbolic groups à la Gromov, written by the two authors in collaboration with T. Delzant, and published as Lecture Notes in Mathematics No. 1441, Springer Verlag 1990. The general theory of hyperbolic groups and spaces have given rise to several publications, each presenting some different aspects of the theory (see in particular the bibliography of Chapter 1). But a systematic presentation of the subject which is treated here had not been considered before (except, of course, for the sketch of the theory which has been given by Gromov). On the other hand, the present volume can be read in a completely self-contained manner. In particular, we have included in Chapter 1 a review of the basic theory of hyperbolic spaces and groups, which is sufficient for the comprehension of the entire text. In the same way, all the necessary notions of the theory of dynamical systems and of symbolic dynamics are included in full detail in Chapter 2.

We would like to thank Thomas Delzant for several suggestions and for his efficient aid in Chapter 4.

Chapter 1

A quick review of Gromov hyperbolic spaces

This chapter is a review of a few basic points in the theory of hyperbolic metric spaces in the sense of Gromov. The main reference for this theory is Gromov's paper [Gro 3], but we can also find an outline of parts of that theory in previous papers of Gromov, mainly [Gro 1] and [Gro 2].

Our main objective in this chapter is to collect, for the purpose of easy reference later in this text, all the statements of the general theory which will be useful for us. Most often, these statements shall be given without proof. (For detailed proofs, the reader is referred to the following papers, listed in the bibliography: [Bow], [CDP], [GH], [Ghy], [Gro 3] and [Sho].) A reader who is familiar with Gromov's theory may very well skip this chapter.

§1 – Hyperbolic metric spaces

Let X be a metric space. We denote by $\mid x - y \mid$ the distance between the points x and y of X.

A *geodesic* in X is a map $\sigma : I \to X$ defined on an interval I of $\mathbb{R}$ such that

$$\mid \sigma(t_1) - \sigma(t_2) \mid = \mid t_1 - t_2 \mid$$

for every $t_1, t_2 \in I$. We shall often identify a geodesic with its image. If $I = [a, b]$ (resp. $[0, \infty[$), we say that the geodesic $\sigma : I \to X$ is a *geodesic segment* (resp. a *geodesic ray*). We denote by $[x, y]$ *any* geodesic segment joining the two points x and y of X. The space X is said to be *geodesic* if any two points in X can be joined by a geodesic segment. For example, any complete Riemannian manifold is geodesic, by the theorem of Hopf-Rinow.

If X is equipped with a basepoint x_0, we shall use the notation

$$\mid x \mid = \mid x_0 - x \mid$$

for every x in X. The *Gromov product* of the points x and y of X is the nonnegative real number $(x.y)$ defined by the formula

$$(x.y) = (x.y)_{x_0} = \frac{1}{2}(\mid x \mid + \mid y \mid - \mid x - y \mid).$$

Definition 1.1. — Let δ be a nonnegative real number. The metric space X is said to be δ-*hyperbolic* if

$$(x.y) \geq min((x.z), (y.z)) - \delta$$

for every $x, y, z \in X$ and for every choice of a basepoint x_0.

Definition 1.2. — The metric space X is said to be *hyperbolic* if there exists a real number δ such that X is δ-hyperbolic.

Examples.
1) Every bounded metric space X is δ-hyperbolic with $\delta = diam(X)$.
2) Every real tree is 0-hyperbolic. Recall that a *real tree* is a geodesic space X which satisfies the following property: given two arbitrary points x and y in X, there exists a unique topological segment joining these points. (Recal that a *topological segment* in a topological space is a subset which is homeomorphic to a segment $[a, b]$ of $\mathbb{R}$.) Conversely, one can show that any geodesic 0-hyperbolic space is a real tree.
3) The usual hyperbolic space, $\mathbf{H}^n$, is δ-hyperbolic with $\delta = Log\ 3$. More generally, every complete simply connected Riemannian manifold whose sectional curvature is everywhere $\leq -a^2$, where a is any positive real, is δ-hyperbolic with $\delta = \frac{Log\ 3}{a}$.

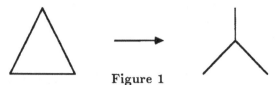

Figure 1

An *n-sided geodesic polygon* in the metric space X is defined by giving n points $x_1, ..., x_n$ in X (the *vertices* of the polygon) and n geodesic segments $[x_1, x_2]$, ..., $[x_{n-1}, x_n]$, $[x_n, x_1]$ (the *sides* of the polygon). By abuse of language, we shall denote by $[x_1, ..., x_n]$ any geodesic polygon with vertices $x_1, ..., x_n$. A geodesic polygon is said to be *ϵ-narrow* if every point situated on one of its sides is at distance $\leq \epsilon$ from the union of the other sides. We have the following

Proposition 1.3. — *In a geodesic δ-hyperbolic metric space, every geodesic triangle is 4δ-narrow.* ∎

Let us note that conversely, a geodesic space in which every geodesic triangle is δ-narrow is δ-hyperbolic. By joining with geodesics one of the vertices of an n-sided geodesic polygon to the $n-3$ vertices which are not adjacent to that vertex (an operation which subdivises the polygon into $n-2$ geodesic triangles), the preceding proposition gives immediately the following

Corollary 1.4. — *In a geodesic δ-hyperbolic space, every n-sided geodesic polygon $(n \geq 3)$ is $4(n-2)\delta$-narrow.* ∎

Remark. In fact, there is a more judicious manner of subdividing a polygon, which shows that under the hypotheses of the preceding corollary, the polygon is $f(n)\delta$-narrow, where $f(n)$ is a function of n which has logarithmic growth (*cf.* [CDP], Lemma 1.5 of chapter 3), but as we shall only apply this result with small values of n, the statement which we give is sufficient.

In a geodesic hyperbolic space, the Gromov product of two points is uniformly close to the distance to the basepoint of any geodesic segment joining these two points. More precisely, we have the

Proposition 1.5. — *Let X be a geodesic space which is δ-hyperbolic and equipped with a basepoint x_0 and let x and y be two points of X. Then,*

$$dist\big(x_0, [x,y]\big) - 4\delta \leq (x.y) \leq dist\big(x_0, [x,y]\big)$$

for every geodesic segment $[x, y]$ joining x and y. ∎

Given a geodesic triangle $\Delta = [x_1, x_2, x_3]$, we associate to it in a natural manner a *tripode*, which we denote by T_Δ (this is the matric graph represented in figure 1), and a map $f_\Delta : \Delta \to T_\Delta$ whose restriction to each side of Δ is distance-preserving. The three points of Δ whose image by f_Δ is equal to the central point of the tripode are called the *internal points* of the triangle.

We say that the triangle Δ is δ-*thin* if any two points of Δ which have the same image by f_Δ are at distance $\leq \delta$ from each other.

Proposition 1.6. — *Let X be a geodesic metric space.*

(i) *If X is δ-hyperbolic, then every geodesic triangle in X is 4δ-thin.*

(ii) *If every geodesic triangle in X is δ-thin, then X is δ-hyperbolic.* ∎

Let us note another way of formulating hyperbolicity for metric spaces, which does not involve basepoints:

Proposition 1.7. — *A metric space X is δ-hyperbolic if and only if we have*

$$|\,x-y\,|+|\,z-t\,| \leq max\,(\,|\,x-z\,|+|\,y-t\,|,|\,x-t\,|+|\,y-z\,|\,)\,+2\delta$$

for all x,y,z and t in X. ∎

§2 – Hyperbolic groups

Consider a group Γ, and let $G \subset \Gamma$ be a generating set of this group. For all $\gamma_1, \gamma_2 \in \Gamma$, let

$$|\,\gamma_1 - \gamma_2\,|_G = min\{n \mid \gamma_1^{-1}\gamma_2 = a_1 a_2 ... a_n,\ a_i \in G \cup G^{-1}\}.$$

This defines a metric $|\,\,|_G$ on Γ, which is called the *word metric* relative to G. Suppose now that Γ is finitely generated, and that G is a finite generating set. We say that Γ is *hyperbolic* if the metric space $(\Gamma, |\,\,|_G)$ is hyperbolic. It can be shown (see §6) that this definition does not depend on the choice of the finite generating set G of Γ.

Examples.
1) Every finite group is hyperbolic.
2) Every free group of finite rank is hyperbolic.
3) Every group which admits a finite presentation which has the small cancellation property $C'(1/6)$ is hyperbolic.
4) The free product of two hyperbolic groups is hyperbolic.
5) Let Γ be a group and let Γ' be a finite-index subgroup of Γ. Then Γ' is hyperbolic if and only if Γ is hyperbolic.

Geometric considerations give access to various properties of hyperbolic groups. Here are some of these properties:

1) A hyperbolic group is finitely presented.

2) A hyperbolic group contains only finitely many conjugacy classes of torsion elements.

3) A hyperbolic group cannot contain a subgroup isomorphic to $\mathbb{Z} \oplus \mathbb{Z}$.

4) A hyperbolic group is automatic in the sense of [E].

5) The growth function of a hyperbolic group, relatively to an arbitrary finite generating set, is rational. (Let us recall that the *growth function* of a group Γ, relatively to a finite generating subset G, is the formal power series $f(t) = \sum c_n t^n$, where c_n is the number of elements of Γ at distance n from the identity element. (Distances are measured in the word metric associated to G.)

§3 – The boundary of a hyperbolic space

Let X be a hyperbolic space. There is a a boundary which is associated to X, and which is defined in the following manner. First, choose a basepoint x_0 of X. We say that a sequence (x_n) of points in X *converges at infinity* if the Gromov product $(x_n.x_p)$ tends to ∞ when (n,p) converges to ∞. It is easy to see that this definition does not depend on the choice of x_0. We then define the following relation $\mathcal{R}$ on the set of sequences of points of X:

$$(x_n)\mathcal{R}(y_n) \iff (x_n.y_n) \to \infty \text{ when } n \to \infty.$$

(Let us note that if we choose, for each n, a geodesic segment $[x_n, y_n]$ joining the points x_n and y_n, then we have $(x_n)\mathcal{R}(y_n)$ if and only if the distance from the basepoint to the segment $[x_n, y_n]$ tends to ∞ when n tends to ∞. This is a consequence of Proposition 1.5.)

The restriction of $\mathcal{R}$ to the set of sequences that converge at infinity is an equivalence relation whose quotient set is, by definition, the *(hyperbolic) boundary* of X. We denote by ∂X this boundary of X. If ξ is a point in ∂X, we say that a sequence of points in X *converges to* ξ if this sequence belongs to ξ. (This definition does not depend on the choice of a basepoint and is consistent with the topology which we shall define later on, on the union $X \cup \partial X$.) The following proposition is an easy consequence of the definitions.

Proposition 3.1. — *Let (x_n) and (y_n) be sequences of points in X. Suppose that the sequence (x_n) converges to a point ξ of ∂X and that there exists a real number C such that $\mid x_n - y_n \mid \leq C$ for every n. Then the sequence (y_n) converges also to ξ.* ∎

If $r : [0, \infty[\to X$ is a geodesic ray, then there exists a point ξ in ∂X such that $r(t_n)$ converges to ξ for every sequence (t_n) of real numbers ≥ 0 such that $t_n \to \infty$. We shall write $\xi = r(\infty)$. In the same manner, every geodesic $\gamma : \mathbb{R} \to X$ defines two distinct points $\gamma(-\infty)$ and $\gamma(\infty)$ of ∂X.

We extend the definition of a geodesic polygon, given in §1, by allowing certain vertices of the polygon to belong to ∂X. An *n-sided geodesic polygon* $\Pi = [x_1, ..., x_n]$ is therefore given by n points $x_1, ..., x_n \in X \cup \partial X$ (the *vertices* of Π) and n geodesics

$\gamma_1, ..., \gamma_n$ (the *sides* of Π) with γ_i joining x_i and x_{i+1} for each i mod n. Let us recall that the geodesic polygon Π is said to be ϵ-*narrow* if every side of Π is contained in the ϵ-neighborhood of the union of the other sides. The following proposition generalizes Corollary 1.4.

Proposition 3.2. — *In a δ-hyperbolic geodesic space X, every n-sided geodesic polygon with p vertices in ∂X and $n - p$ vertices in X $(n + p \geq 3)$ is $4(n + p - 2)\delta$-narrow.*

PROOF. Let $\Pi = [x_1, x_2, ..., x_n]$ be such a polygon. We begin by "truncating" Π at each vertex situated on the boundary in the following manner: If $x_i \in \partial X$, we place a point y_i (resp. z_i) on the side which goes from x_{i-1} to x_i (resp. from x_i to x_{i+1}), the indices being taken modulo n. We then replace, in the polygon Π, $[y_i, x_i] \cup [x_i, z_i]$ by a geodesic segment $[y_i, z_i]$. We thus obtain an $(n + p)$-sided polygon, with has all its vertices in X. Let x be a point situated on one of the sides γ of Π, and let us take x as a basepoint. We can then place the points y_i and z_i sufficiently close to the point x_i so that $(y_i.z_i) > 4(n + p - 2)\delta$, for every i such that $x_i \in \partial X$. Corollary 1.4 and Proposition 1.5 show then that the point x is at a distance $\leq 4(n + p - 2)\delta$ from the union of the sides of Π other than γ. ∎

Corollary 3.3. — *Let $\gamma_1, \gamma_2 : \mathbb{R} \to X$ be geodesics such that $\gamma_1(-\infty) = \gamma_2(-\infty)$ and $\gamma_1(\infty) = \gamma_2(\infty)$. Then every point on γ_1 is at distance $\leq 8\delta$ from γ_2.* ∎

Corollary 3.4. — *Let $r_1, r_2 : [0, \infty[\to X$ be geodesic rays such that $r_1(0) = r_2(0)$ and $r_1(\infty) = r_2(\infty)$. Then every point on r_1 is at distance $\leq 4\delta$ from r_2.* ∎

Remark. In fact, we can see, using the fact that the triangle $[r_1(0), r_1(T), r_2(T)]$ is 4δ-thin and taking T large enough, that the folowing inequality is satisfied, for every $t \geq 0$:

$$(3.4.1) \quad \mid r_1(t) - r_2(t) \mid \leq 4\delta.$$

Corollary 3.5. — *Let $r_1, r_2 : [0, \infty[\to X$ be geodesic rays such that $r_1(\infty) = r_2(\infty)$. Then every point on r_1 is at distance $\leq \mid r_1(0) - r_2(0) \mid +8\delta$ from r_2. Furthermore, there exists a real number $T \geq 0$ such that $r_1(t)$ is at distance $\leq 8\delta$ from r_2, for every $t \geq T$.* ∎

Recall that a metric space is said to be *proper* if all its bounded subsets are relatively compact. Recall also that a geodesic space is proper if and only if it is complete and locally compact (see for instance [GLP], Theorem 1.10).

In the rest of this paragraph, we suppose that the space X is geodesic, δ-hyperbolic and proper.

Under these hypotheses, Ascoli's theorem shows that given a point in X and a point in ∂X (resp. two distinct points in ∂X), there exists a geodesic ray (resp. a geodesic) joining these points.

The topology on $X \cup \partial X$ is defined in the following manner.

Let us take a basepoint x_0 for X. Let R be the set of geodesics $\sigma : I \to X$ such that I is an interval of type $[0, T]$ (where T is an arbitrary nonnegative real number) or the interval $[0, \infty[$ with $\sigma(0) = x_0$. We extend every $\sigma \in R$, defined on an interval $[0, T]$, to the interval $[0, \infty[$ by taking $\sigma(t) = \sigma(T)$ for every $t \geq T$. The set R appears then as an equicontinuous set of maps from $[0, \infty[$ to X. We equip R with the topology of uniform convergence on compact sets. The compactness of R is a consequence of Ascoli's theorem. There is a natural surjective map $\pi : R \to X \cup \partial X$ defined by $\pi(\sigma) = \sigma(\infty)$ for $\sigma \in R$. We equip $X \cup \partial X$ with the quotient topology. It can be shown that this topology does not depend on the choice of the basepoint x_0 and that the topology induced on X is compatible with the metric. $X \cup \partial X$ is a compact set in which X is open and dense. Every isometry of X can be extended into a homeomorphism of $X \cup \partial X$.

Examples.

1) If X is a real tree which is proper, the map which associates to r the point $r(\infty)$ is a homeomorphism from the set of geodesic rays starting at the basepoint x_0, equipped with the topology of uniform convergence on the compact sets of $[0, \infty[$, to ∂X.

2) Let X be a complete and simply connected Ricmannian manifold whose sectional curvature is bounded above by a negative constant. Let $S(x_0)$ be the unit tangent sphere of X at x_0. For every $v \in S(x_0)$, let r_v be the geodesic ray in X such that $r'_v(0) = v$. Then the map which associates to each such v the point $r_v(\infty)$ is a homeomorphism from $S(x_0)$ to ∂X.

Let us recall that the Gromov product $(\ . \ ) : X \times X \to \mathbb{R}$ can be naturally extended to the whole space $(X \cup \partial X) \times (X \cup \partial X)$. This extension is defined in the following way:

If $a = (a_i)_{i \geq 0}$ and $b = (b_i)_{i \geq 0}$ are two sequences of points in X, we let

$$(a.b) = lim \ inf \ (a_i.b_i) \, when \ i \to \infty.$$

If x and y are two points of $X \cup \partial X$, we let

$$(x.y) = inf(a.b),$$

with $a = (a_i)$ converging to x and $b = (b_i)$ converging to y. (Note that the formula is correct if the points x and y belong to X).

We then have $(x.y) = +\infty$ if and only if $x = y \in \partial X$, and we still have (by taking limits)

$$(x.y) \geq \min \big((x.z), (y.z) \big) - \delta,$$

for every $x, y, z \in X \cup \partial X$.

§4 – The visual metric on the boundary

In this section, X is a δ-hyperbolic space which is geodesic and proper and equipped with a basepoint x_0.

Let a be a real number which is > 1. For a sufficiently close to 1, Gromov defines (§§7.2.K, 7.2.L and 7.2.M of [Gro 3]), a *visual metric* $|\ |_a$ on ∂X in the following manner. Let us define the *a-length* $\ell_a(\sigma)$ of a continuous path $\sigma : [t_1, t_2] \to X$ (t_1, t_2 real numbers) as the integral (defined by taking limits of "Riemann sums") of the function $f : X \to \mathbb{R}$ defined by $f(x) = a^{-|x|}$, along the path σ. Let us then define the a-distance $|\ x - y\ |_a$ between two arbitrary points x and y of X by

$$|\ x - y\ |_a = \inf \ell_a(\sigma),$$

the lower bound being taken on the set of continuous paths σ joining the points x and y. One can show that there exists a constant $a_0 = a_0(\delta)$ (*i.e.* depending only on δ), with $1 < a_0 \leq \infty$, such that for every real number a which is strictly contained between 1 and a_0, the following properties are satisfied:

$(P1)$ The identity map of X extends to a homeomorphism from $X \cup \partial X$ to the completion of X with respect to the metric $|\ |_a$. In particular, the metric $|\ |_a$ induces a metric on ∂X.

$(P2)$ For all distinct points ξ and η of ∂X and for every geodesic $\gamma : \mathbb{R} \to X$ joining ξ and η, we have, denoting by d the distance from the basepoint x_0 to (the image of) γ :

$$\lambda^{-1} a^{-d} \leq |\ \xi - \eta\ |_a \leq \lambda a^{-d}$$

where $\lambda = \lambda(\delta, a) \geq 1$ is a constant which depends only on δ and on a.

For $a \in]1, a_0(\delta)[$, we call the metric $|\ |_a$ on ∂X *the visual metric with parameter a and basepoint x_0.* Property $(P2)$ emphasizes the "visual" character of such a metric.

Example. Suppose that X is a tree (*i.e.* $\delta = 0$). Let us note that the extension of the Gromov product, $(\ .\ ) : (X \cup \partial X) \times (X \cup \partial X) \to [0, \infty]$ which we have recalled in §3 above admits in that case a very simple geometric interpretation: for ξ and η in ∂X, (ξ, η) is the length of the common path of the two geodesic rays starting at x_0 and ending respectively at ξ and η.

Proposition 4.1. — *If X is a tree, we can take $a_0 = \infty$ and we have:*

$$|\ \xi - \eta\ |_a = (2/Log a) a^{-(\xi.\eta)}$$

for all points ξ and η in ∂X and for every $a > 1$.

PROOF. Let x and y be points in X. Let σ_0 be the geodesic segment joining x and y. It is clear that we have $\ell_a(\sigma) \geq \ell_a(\sigma_0)$ for every continuous path σ joining x and y. Hence, we have

$$|\ x - y\ |_a = \ell_a(\sigma_0).$$

12

By integrating, we find that

$$\ell_a(\sigma_0) = (1/\text{Log } a)(2a^{-(x \cdot y)} - a^{-|x|} - a^{-|y|}).$$

∎

§5 – Approximation by trees

In this section, X is a geodesic δ-hyperbolic space, equipped with a basepoint x_0.

The following theorem (*Approximation by trees*) says that the union of a finite number of geodesics starting at x_0 (that is, a "star" centered at x_0) is, from a metric point of view, uniformly close to a tree (the "uniform" constant depending only on δ and on the number of geodesics involved). This theorem is useful in general hyperbolic spaces for proving inequalities which involve distances between a finite number of points.

Theorem 5.1. — (Approximation by trees). *Let $\gamma_1, ... \gamma_n$ be geodesics in X. Suppose that, for all $i = 1, ..., n$, γ_i is defined on an interval of the form $[0, T]$ or $[0, \infty[$ and satisfies $\gamma_i(0) = x_0$. Let Z be the union of (the images of) the γ_i. Then there exists a real tree with a natural basepoint, (T, t_0), and a map $f : (Z, x_0) \to (T, t_0)$ satisfying the following two properties:*

(1)   *For every z and z' in Z, we have*

$$\mid z - z' \mid -c_n \delta \leq \mid f(z) - f(z') \mid \leq \mid z - z' \mid,$$

where $c_n = 2(1 + log_2 n)$.

(2)   *The restriction of f to (the image of) γ_i is distance preserving, for all i.*   ∎

We shall apply sometimes Theorem 5.1 with a finite set of points $F \subset X \cup \partial X$ instead of the finite set of geodesics. This means that, among these points, there is a basepoint of X which we shall specify, and that we apply the theorem to a set of geodesics γ_i joining the basepoint to each of these points in F.

§6 – Quasi-geodesics and quasi-isometries

Let X and Y be two metric spaces. Given two real numbers $\lambda \geq 1$ and $k \geq 0$, we say that a map $f : X \to Y$ is a (λ, k)-*quasi-isometry* if

$$\lambda^{-1} \mid x_1 - x_2 \mid -k \leq \mid f(x_1) - f(x_2) \mid \leq \lambda \mid x_1 - x_2 \mid +k$$

for every x_1 and x_2 in X. (Note that such a map f is not necessarily continuous.)

In a metric space X, a (λ, k)-*quasi-geodesic* is a (λ, k)-quasi-isometry $\sigma : I \to X$, where I is an interval of $\mathbb{R}$ or of $\mathbb{Z}$. In the case where $I = \mathbb{N}$, we say that we have a (λ, k)-*quasi-geodesic* sequence.

When is not necessary to specify the values of λ and k, we can simply say *quasi-isometry* (resp. *quasi-geodesic*), instead of (λ, k)-quasi-isometry (resp. (λ, k)-quasi-geodesic).

The following theorem is fundamental in the theory of hyperbolic spaces.

Theorem 6.1. — (Stability of quasi-geodesics). *Let X be a geodesic space which is δ-hyperbolic. Let I be a segment in $\mathbb{R}$ or in $\mathbb{Z}$, whose endpoints are a and b, and $\sigma : I \to X$ a (λ, k)-quasi-geodesic. Let γ be a geodesic segment in X joining the endpoints $\sigma(a)$ and $\sigma(b)$ of σ. Then the (images of) σ and γ are at a Hausdorff distance $\leq K$ from each other, where $K = K(\delta, \lambda, k)$ is a constant which depends only on δ, λ and k.* ∎

(Let us recall that we say that two subsets of a metric space are at *Hausdorff distance* $\leq \epsilon$ from each other if each of these subsets is contained in the ϵ-neighborhood of the other.)

This Theorem gives immediately the following corollaries:

Corollary 6.2. — *Let $\sigma : [0, \infty[\to X$ (resp. $\sigma : \mathbb{N} \to X$) be a (λ, k)-quasi-geodesic. Then $\sigma(t)$ has a limit $\sigma(\infty) \in \partial X$ when $t \to \infty$. Furthermore, if $r : [0, \infty[\to X$ is a geodesic ray such that $r(0) = \sigma(0)$ and $r(\infty) = \sigma(\infty)$, then (the images of) σ and r are at a Hausdorff distance $\leq K'$ from each other, where $K' = K'(\delta, \lambda, k)$ is a constant which depends only on δ, λ and k.* ∎

Corollary 6.3. — *Let $\sigma : \mathbb{R} \to X$ (resp. $\sigma : \mathbb{Z} \to X$) be a (λ, k)-quasi-geodesic. Then $\sigma(t)$ admits a limit $\sigma(\infty) \in \partial X$ (resp. $\sigma(-\infty) \in \partial X$) when $t \to \infty$ (resp. $t \to -\infty$). We have $\sigma(\infty) \neq \sigma(-\infty)$. Furthermore, if $\gamma : \mathbb{R} \to X$ is a geodesic such that $\sigma(-\infty) = \gamma(-\infty)$ and $\sigma(\infty) = \gamma(\infty)$, then (the images of) σ and γ are at Hausdorff distance $\leq K''$ from each other, where $K'' = K''(\delta, \lambda, k)$ is a constant which depends only on δ, λ and k.* ∎

Corollary 6.4. — *Let X and Y be two geodesic spaces. Suppose that Y is hyperbolic and let $f : X \to Y$ be a quasi-isometry. Then*

(1)   *X is hyperbolic.*

(2)   *For every sequence (x_n) of points in X which converges to a point ξ in ∂X, the sequence $(f(x_n))$ converges to a point of ∂Y which depends only on ξ, and which we denote by $\partial f(\xi)$.*

(3)   *The map $\partial f : \partial X \to \partial Y$ is injective.*

(4)   *If $f(X)$ is ϵ-dense in Y, then the map $\partial f : \partial X \to \partial Y$ is surjective.*

(5)   *If X and Y are proper, then ∂f is a topological imbedding from ∂X into ∂Y.* ∎

Application to hyperbolic groups.

Let Γ be a group and $G \subset \Gamma$ a finite generating set of Γ. Recall that the *Cayley graph* $K(\Gamma, G)$ of Γ relatively to G is the simplicial graph defined in the following manner. The vertices of $K(\Gamma, G)$ are the elements of Γ and two distinct elements γ and γ' in Γ are related by an edge if and only if there exists an element g in G such that $\gamma = \gamma' g$. It follows from the fact that G generates Γ that $K(\Gamma, G)$ is connected. $K(\Gamma, G)$ is equipped with its canonical metric which will be denoted by $|\ |_G$. (The *canonical metric* of a connected simplicial graph K is the maximal metric on K for which every edge is isometric to the interval $[0, 1]$.) One should note that the restriction of $|\ |_G$ to Γ is the word metric relative to G, defined in §2 above. It is easy to verify that $K(\Gamma, G)$ is geodesic and proper. The space $(K(\Gamma, G), |\ |_G)$ is hyperbolic if and only if $(\Gamma, |\ |_G)$ is hyperbolic. If G' is another generating set of Γ, one can show that the identity map of Γ extends to a quasi-isometry $K(\Gamma, G) \to K(\Gamma, G')$. Corollary 6.4 implies therefore that the hyperbolicity of $(\Gamma, |\ |_G)$ depends only on Γ (a result which we have already announced in §2).

Suppose now that Γ is hyperbolic. Corollary 6.4 shows that there is a canonical homeomorphism $\partial K(\Gamma, G) \to \partial K(\Gamma, G')$ which is induced by the identity map of Γ. Let us define $\partial \Gamma = \partial K(\Gamma, G)$. As a topological space, $\partial \Gamma$ is compact and metrizable. Of course, it is called the *boundary* of the hyperbolic group Γ. Let us also note that the action of Γ on $K(\Gamma, G)$ by left translations is an isometric action which induces a (continuous) action of Γ on $\partial \Gamma$. This action on $\partial \Gamma$ does not depend on the choice of the finite generating set G of Γ. The dynamical system $(\partial \Gamma, \Gamma)$ is therefore canonically associated to the hyperbolic group Γ. The study of this dynamical system is the main subject matter of these notes.

Examples.
1) A hyperbolic group has empty boundary if and only if it is a finite group.
2) If $\Gamma = \mathbb{Z}$, then $\partial \Gamma = \{-\infty, \infty\}$.
3) If Γ is a free group of rank n, $n \geq 2$, then $\partial \Gamma$ is a Cantor set, that is, $\partial \Gamma$ is homeomorphic to $\{0, 1\}^{\mathbb{N}}$.

The next statement, which is very useful for proving the hyperbolicity of certain groups and for studying their boundary, is an easy consequence of Corollary 6.4.

Theorem 6.5. — *Let X be a geodesic space which is proper. Let Γ be a group of isometries of X which acts properly discontinuously on this space, and suppose that this action is cocompact. Then X is hyperbolic as a metric space if and only if Γ is hyperbolic. Furthermore, if X is hyperbolic, we have a canonical homeomorphism $\partial \Gamma \to \partial X$ which is Γ-equivariant.* ∎

(Let us recall that a group Γ acts *cocompactly* on a topological space X if the quotient space X/Γ is compact. Recall also that we say that a group Γ acts *properly discontinuously* on a locally compact space X if for every compact subset $K \subset X$, the set of elements γ in Γ such that $\gamma K \cap K \neq \emptyset$ is finite. Finally, recall that if E and F

are sets equipped with an action of Γ, a map $f : E \to F$ is said to be Γ-*equivariant* if $f(\gamma x) = \gamma f(x)$, for every $\gamma \in \Gamma$ and for every $x \in E$).

Example. Consider the space $\mathbf{H}^n$ or, more generally, a complete simply connected Riemannian manifold of dimension n whose sectional curvature is bounded above by a negative constant. The previous theorem shows that any discrete and cocompact subgroup of $Isom(X)$ is hyperbolic, and that its boundary is homeomorphic to S^{n-1}.

§7 – Classification of isometries

In this section, X is again a geodesic metric space, which is δ-hyperbolic and proper.

The isometries of X are classified into three distinct types, according to the behaviour of a point in X under iteration by the given isometry. This classification generalizes the well-known classification of isometries of $\mathbf{H}^n$.

Definition 7.1. — Let γ be an isometry of X and let x be an arbitrary point of X. We say that γ is *elliptic* if the sequence $(\gamma^n x)_{n \in \mathbb{Z}}$ is bounded. We say that γ is *hyperbolic* if the sequence $(\gamma^n x)_{n \in \mathbb{Z}}$ is a quasi-geodesic. Finally, we say that γ is *parabolic* if γ is neither elliptic nor hyperbolic. It can easily be shown that this definition does not depend on the choice of the point $x \in X$.

If γ is a hyperbolic isometry and x a point in X, the sequence $\gamma^n x$ admits a limit $\gamma^+ \in \partial X$ (resp. $\gamma^- \in \partial X$) when n tends to ∞ (resp. $-\infty$) and we have $\gamma^+ \neq \gamma^-$, according to Corollary 6.3. It follows from Proposition 3.1 that the points γ^+ and γ^- do not depend on x. Indeed, if x' is another point of X, we have $| \gamma^n x - \gamma^n x' | = | x - x' |$. It is clear that γ^+ and γ^- are fixed points of γ. The point γ^+ (resp. γ^-) is called the *attracting* (resp. *repelling*) fixed point of γ.

Recall that we have denoted by $| \ |_a$ the visual metric of parameter a on ∂X. Recall also that the map $f : X_1 \to X_2$ between metric spaces is said to be λ-*Lipschitz* if we have $| f(x) - f(y) | \leq \lambda | x - y |$ for every x and y in X_1. We have the following

Proposition 7.2. — *Let γ be a hyperbolic isometry of X, with γ^+ (resp. γ^-) its attracting (resp. repelling) fixed point. $\in \partial X$. Then*

(1) *For every point $\xi \in \partial X - \{\gamma^-\}$ (resp. $\partial X - \{\gamma^+\}$), the sequence $\gamma^n(\xi)$ tends to γ^+ (resp. γ^-) when n tends to ∞ (resp. $-\infty$). Furthermore, the convergence is uniform on every compact subset of $\partial X - \{\gamma^-\}$ (resp. $\partial X - \{\gamma^+\}$).*

(2) *Given two neighborhoods U^+ and U^- in ∂X of the points γ^+ and γ^- respectively, and a real number $\lambda > 1$, there exists an integer n_0 such that γ^n (resp. γ^{-n}) sends $\partial X - U^-$ (resp. $\partial X - U^+$) into U^+ (resp. U^-) and is λ-Lipschitz on $\partial X - U^-$ (resp. $\partial X - U^+$) for all $n \geq n_0$.* ∎

Let γ be a parabolic isometry of X. One can show that γ fixes a unique point on ∂X. Furthermore, we have the following

Proposition 7.3. — *Let γ be a parabolic isometry of X, with $\alpha \in \partial X$ as fixed point. Then*

(1)   *For every point $\xi \in \partial X$, the sequence $\gamma^n(\xi)$ tends to α when $\mid n \mid \to \infty$. Furthermore, the convergence is uniform on compact sets of $\partial X - \{\alpha\}$.*

(2)   *Given a neighborhood U of α in ∂X, there exists an integer n_0 such that γ^n sends $\partial X - U$ into U for every n such that $\mid n \mid \geq n_0$.* ∎

(Note that, in the preceding two propositions, (1) is implied by (2).)

Remarks.
1) A real tree does not have parabolic isometries.
2) Let Γ be a hyperbolic group and X its Cayley graph for a finite generating set $G \subset \Gamma$. Let us recall that Γ acts isometrically on X. If γ is a torsion element of Γ, it is clear that γ defines an elliptic isometry of X. If γ is a non-torsion element of Γ, it can be shown that it defines a hyperbolic isometry of X. (In particular, the type of the isometry of X which is defined by an element of the group Γ does not depend on the choice of the generating set G.)

§8 – The polyhedron $P_d(X)$

Let X be a metric space and let d be a positive real number. We construct a simplicial complex, denoted by $P_d(X)$, in the following manner. The vertices of $P_d(X)$ are the points of X. The simplices of $P_d(X)$ have as a set of vertices the finite subsets of X whose diameter is $\leq d$.

Examples.
1) If X is bounded, then $P_d(X)$ is the standard simplex in $\mathbb{R}^X$, provided $d \geq diam(X)$.
2) Let Γ be a group, and G a generating set of Γ. We equip Γ with the word metric relative to G. Then $P_1(\Gamma)$ is the Cayley graph $K(\Gamma, G)$, as it has been defined in §6.

The complex $P_d(X)$ is a sort of "regularizing space" for the metric space X. An important phenomenon, when X is hyperbolic, is the stable contractibility of $P_d(X)$. More precisely, we have the following theorem (due to Rips):

Theorem 8.1. — *Let X be a geodesic δ-hyperbolic space, and let d be a positive real number. Then the simplicial complex $P_d(X)$ is contractible for all $d \geq 4\delta$.* ∎

Bibliography for Chapter 1

[Bow] B. Bowditch, "Notes on Gromov's hyperbolicity criterion for path-metric spaces", *in Group Theory from a geometrical viewpoint*, ICTP, World Scientific, 1991, pp. 64–167.

[CDP] M. Coornaert, T. Delzant and A. Papadopoulos, "Géométrie et théorie des groupes: Les groupes hyperboliques de Gromov", Lecture Notes in Mathematics, vol. 1441, Springer-Verlag, 1990.

[E] D. B. A. Epstein (with J. W. Cannon, D. F. Holt, S. V. F. Levy, M. S. Paterson, and W. P. Thurston), "Word processing and groups", Jones and Barnett Publishers, 1992.

[GH] E. Ghys, P. de la Harpe (ed), "Sur les groupes hyperboliques d'après Mikhaël Gromov", Progress in Mathematics, vol. 83, Birkhäuser, 1990.

[Ghy] E. Ghys, "Les groupes hyperboliques", Seminaire N. Bourbaki, exposé No. 772, mars 1990. Astérisque **189-190**, SMF, 1990.

[GLP] M. Gromov, "Structures métriques pour les variétes riemanniennes", notes written by J. Lafontaine and P. Pansu, Fernand Nathan, Paris, 1981.

[Gro 1] ——, "Hyperbolic manifolds, groups and actions", *in Riemann surfaces and related topics*, Ann. of Math. studies **97**, Princeton University Press, 1980, pp. 183–213.

[Gro 2] ——, "Infinite groups as geometric objects", Proc. ICM Warszawa, 1983, pp. 385–392.

[Gro 3] ——, "Hyperbolic groups", in *Essays in group theory*, MSRI publ. 8, Springer Verlag, 1987, pp. 75–263.

[Sho] H. Short (ed.), "Notes on word hyperbolic groups", *in Group Theory from a geometrical viewpoint*, ICTP, World Scientific, 1991, pp. 3–63.

Chapter 2

Symbolic dynamics

Consider a finite set S, which will be called the set of *symbols*, and let Γ be a countable semigroup. (Recall that a *semigroup* is a set equipped with an internal law which is associative and with an identity element.)

Symbolic dynamics studies the action of Γ on the set Σ of maps $\sigma : \Gamma \to S$. The action of Γ on Σ is given by $\gamma\sigma(\gamma') = \sigma(\gamma'\gamma)$ ($\sigma \in \Sigma$ and $\gamma, \gamma' \in \Gamma$). We equip Γ and S with the discrete topology and Σ with the product topology. The dynamical system (Σ, Γ) is called the *Bernoulli shift* associated to Γ and S.

Using methods which are essentially of combinatorial nature, symbolic dynamics is useful for understanding the behaviour of certain dynamical systems (Ω, Γ), where Ω is a space on which Γ acts. The general idea consists in "coding" in an equivariant way the elements of Ω by elements of some Bernoulli shift Σ (with an appropriate set of symbols S).

This symbolic approach has been adopted (with $\Gamma = \mathbb{Z}$) in the study of Anosov diffeomorphisms (Sinai) and, more generally, in the study of Axiom A diffeomorphisms (Smale, Bowen, Manning,...). One of the results which have been obtained is the rationality of the ζ-function for Axiom A diffeomorphisms, which has been proved by Manning ([Man]). A paper of Hadamard ([Had]), published in 1898, inaugurates a long series of works concerning the application of symbolic dynamics to the study of geodesic flows in negative curvature. This approach has been adopted by several mathematicians (Koebe, Nielsen, Morse, Hedlund, Gromov, Series,...).

In this chapter, we present, following Gromov ([Gro 1] and [Gro 3], see also Fried [Fri]), a few basic notions of symbolic dynamics. In particular, we introduce the notions of *dynamical systems of finite type* and of *finitely presented dynamical systems*. (The reader will notice a certain analogy between the terminology used for the theory of symbolic dynamics and the theory of groups. This analogy, which is stressed by Gromov, will be apparent as the theory will go on (*cf.* [Fri]).)

§1 – Bernoulli shifts

Let Γ be a countable semigroup and S a finite set. The elements of S are called the *symbols*. We denote by $\Sigma = \Sigma(\Gamma, S)$ the set S^Γ of maps $\sigma : \Gamma \to S$. We equip Γ and S with the discrete topology and Σ with the product topology. Thus, a sequence (σ_n) of elements of Σ converges to an element $\sigma \in \Sigma$ if and only if for every $\gamma \in \Gamma$, there exists an integer $n_0 = n_0(\gamma)$ such that $\sigma_n(\gamma) = \sigma(\gamma)$ for every $n \geq n_0$.

Proposition 1.1. — *The space Σ is metrizable, compact and totally disconnected. Furthermore, if Γ is infinite and if $card(S) \geq 2$, then Σ is a perfect set.*

(Let us recall that a topological space is said to be *perfect* if it has no isolated points, and *totally disconnected* if each of its connected components is reduced to a point.)

PROOF. If $| \ |_S$ is an arbitrary metric on S and $(\alpha_\gamma)_{\gamma \in \Gamma}$ a family of positive real numbers such that $\sum \alpha_\gamma < \infty$, it is clear that the metric $| \ |$ on Σ defined by

$$(1.1.1) \qquad | \sigma - \sigma' | = \sum_{\gamma \in \Gamma} \alpha_\gamma \, | \sigma(\gamma) - \sigma'(\gamma) |_S$$

is compatible with the topology of Σ. The compactness of Σ is a consequence of Tychonoff's theorem (by a diagonal process, we extract from each sequence in Σ a convergent subsequence). To show that Σ is totally disconnected, consider the map $p_\gamma : \Sigma \to S$ which associates to $\sigma \in \Sigma$ the element $\sigma(\gamma)$. This map is continuous and therefore $p_\gamma(C)$ is reduced to a point for every connected non-empty set $C \subset \Sigma$. Suppose now that Γ is infinite and let us show that in this case Σ is perfect if $card(S) \geq 2$. Given $\sigma \in \Sigma$, we construct a sequence (σ_n) of distinct elements of Σ which converges to σ, in the following manner. Let (γ_n) be a sequence of distinct elements of Γ. Define $\sigma_n : \Gamma \to S$ by the formula $\sigma_n(\gamma) = \sigma(\gamma)$ if $\gamma \neq \gamma_n$ and $\sigma_n(\gamma_n)$ = an arbitrary element of S which is distinct from $\sigma(\gamma)$. It is clear that this sequence σ_n does the job. ∎

Corollary 1.2. — *If Γ is infinite and $card(S) \geq 2$, then Σ is a Cantor set.*

(Recall that a *Cantor set* is a topological space which is homeomorphic to the triadic Cantor set, that is, homeomorphic to the space $\{0, 1\}^N$ equipped with the product topology.)

PROOF. From a classical theorem of general topology (see, for instance, [Moi], Chapter 12, Theorem 8) any compact, metrizable, perfect and totally disconnected space is a Cantor set. ∎

We define now a continuous left action of Γ on Σ. For $\gamma \in \Gamma$ and $\sigma \in \Sigma$, $\gamma\sigma$ is defined by the following rule:

$$\forall \gamma' \in \Gamma, \gamma\sigma(\gamma') = \sigma(\gamma'\gamma).$$

The space Σ, equipped with this action of Γ, is called the *Bernoulli shift* on Γ (and with associated set of symbols S) .

Exercise. Let Γ be an infinite countable semigroup, S a finite set and $\Sigma = \Sigma(\Gamma, S)$. Assume that the following condition holds:

(*) For any finite subset $F \subset \Gamma$, there exists an element $\gamma \in \Gamma$ such that $F\gamma \cap F = \emptyset$.

(Note that (*) is satisfied if Γ is a group or, more generally, if Γ can be imbedded in some group.)
1) Show that the action of Γ on Σ is topologically transitive (*i.e.* admits a dense orbit).
2) Show that, for any integer $n \geq 1$, the diagonal action of Γ on Σ^n is topologically transitive. (Hint: use $\Sigma^n = \Sigma(\Gamma, S^n)$.)

Definition 1.3. — A *subshift* of Σ is closed and Γ-invariant subset of Σ.

Exercise. Show that the intersection of an arbitrary family of subshifts of Σ is again a subshift of Σ. Show also that the union of a finite family of subshifts of Σ is also a subshift of Σ.

Let now S_1 and S_2 be two finite sets and let us consider the two Bernoulli shifts $\Sigma_1 = \Sigma(\Gamma, S_1)$ and $\Sigma_2 = \Sigma(\Gamma, S_2)$. Given a finite subset $F \subset \Gamma$ and a map $u : S_1^F \to S_2$, we construct a continuous and Γ-equivariant map $u_\infty : \Sigma_1 \to \Sigma_2$ in the following way. For every $\sigma \in \Sigma_1$, we define $u_\infty(\sigma) \in \Sigma_2$ by the formula:

$$\forall \gamma \in \Gamma, \ u_\infty(\sigma)(\gamma) = u(\gamma\sigma_{|F}).$$

The continuity and the equivariance of u_∞ can be verified immediately. In fact, one can obtain in this way all the continuous and equivariant maps between subshifts. More precisely, we have the following

Proposition 1.4. — *Let $\Phi_1 \subset \Sigma(\Gamma, S_1)$ and $\Phi_2 \subset \Sigma(\Gamma, S_2)$ be two subshifts. If $T : \Phi_1 \to \Phi_2$ is continuous and Γ-equivariant, then there exists a finite subset $F \subset \Gamma$ and a map $u : S_1^F \to S_2$ such that T is the restriction to Φ_1 of the map u_∞ defined above.*

PROOF. By compactness, the map T is uniformly continuous on Φ_1. Therefore, we can find an $\epsilon > 0$ such that if the elements $\sigma_1, \sigma_2 \in Phi_1$ are at distance less than ϵ (as defined by formula (1.1.1)), then $T(\sigma_1)$ and $T(\sigma_2)$ take the same value at the identity element $Id \in \Gamma$. Hence, there is a finite set $F \subset \Gamma$ such that $T(\sigma_1) = T(\sigma_2)$ whenever σ_1 and σ_2 have the same restriction on F.

Thus, we can construct a map $u : S_1^F \to S_2$ such that $u(\sigma_{|F}) = T(\sigma)(Id)$ for every $\sigma \in \Phi_1$. We have, for every $\sigma \in \Phi_1$ and for every $\gamma \in \Gamma$,

$$u_\infty(\sigma)(\gamma) = u(\gamma\sigma_{|F})$$

$$= T(\gamma\sigma)(Id)$$

21

$$= \gamma T(\sigma)(Id)$$

$$= T(\sigma)(\gamma),$$

and therefore

$$u_\infty = T.$$

∎

Corollary 1.5. — *The set of continuous and Γ-equivariant maps $T : \Phi_1 \to \Phi_2$ is countable.* ∎

§2 – Expansive systems

We shall talk of the *dynamical system* (or simply of the *system*) (Ω, Γ) to designate a metrizable compact space Ω equipped of a continuous left action of the semigroup Γ.

Given a dynamical system (Ω, Γ), a subset A of Ω and an element γ of Γ, we define as usual $\gamma^{-1}A$ by

$$\gamma^{-1}A = \{x \in \Omega | \gamma x \in A\}.$$

(Note that γ is not necessarily invertible.)

Definition 2.1. — We shall say that the system (Ω, Γ) is *expansive* if there exists an open set U in $\Omega \times \Omega$ such that $\Delta = \cap_{\gamma \in \Gamma} \gamma^{-1}U$, where

$$\Delta = \{(x, x) \mid x \in \Omega\} \subset \Omega \times \Omega$$

denotes the diagonal of $\Omega \times \Omega$, and where the action of Γ on $\Omega \times \Omega$ is defined by $\gamma(x, y) = (\gamma x, \gamma y)$, for $\gamma \in \Gamma$ and $x, y \in \Omega$.

Definition 2.2. — Given a system (Ω, Γ) and a metric $| \ |$ on Ω which is compatible with the topology of this space, we say that the real number $\epsilon > 0$ is an *expansivity constant* for $(\Omega, | \ |, \Gamma)$ if for all distinct points x and y in Ω, there exists an element γ in Γ such that $| \gamma x - \gamma y | \geq \epsilon$.

Proposition 2.3. — *Let $| \ |$ be a metric on Ω which is compatible with the topology. Then the system (Ω, Γ) is expansive if and only if there exists an expansivity constant $\epsilon > 0$ for $(\Omega, | \ |, \Gamma)$.*

PROOF. Let ϵ be an expansivity constant for $(\Omega, | \ |, \Gamma)$. Let us define

$$U = \{(x, y) \in \Omega \times \Omega \text{ such that } | x - y | < \epsilon\}.$$

From this definition, we can see easily that U is an open set and satisfies $\Delta = \cap_{\gamma \in \Gamma} \gamma^{-1}U$.

Conversely, suppose that there exists an open set U of $\Omega \times \Omega$ such that $\Delta = \cap_{\gamma \in \Gamma} \gamma^{-1} U$. Define $\epsilon = inf\{|\, x - y \,|$ such that $(x, y) \notin U\}$. We have $\epsilon > 0$, because Δ is a subset of U and the complement of U in $\Omega \times \Omega$ is compact. This ϵ is an expansivity constant for $(\Omega, |\,|, \Gamma)$.   ∎

Examples.
1) If (Ω, Γ) is expansive and if Ω' is a closed Γ-invariant subset of Ω, then (Ω', Γ) is expansive.
2) If S is a finite set and $\Sigma = \Sigma(\Gamma, S)$, then the Bernoulli shift (Σ, Γ) is expansive. Indeed, if σ and σ' are two distinct elements of Σ, there exists an element γ in Γ such that $\sigma(\gamma) \neq \sigma'(\gamma)$. We have therefore $\gamma\sigma(Id) \neq \gamma\sigma'(Id)$ which implies, by taking the metric $|\,|$ defined by formula (1.1.1), $|\, \gamma\sigma - \gamma\sigma' \,| \geq \epsilon > 0$ with

$$\epsilon = \alpha_{Id} inf_{s \neq s'} |\, s - s' \,|_S \,.$$

3) If the dynamical systems (Ω_1, Γ_1) and (Ω_2, Γ_2) are expansive, then the system $(\Omega_1 \times \Omega_2, \Gamma_1 \times \Gamma_2)$ is also expansive.
4) Suppose that there exists a metric $|\,|$ on Ω, compatible with the topology, such that Γ acts isometrically on Ω. Then (Ω, Γ) is not expansive, unless Ω is finite. This implies that any expansive system has only finitely many fixed points.

The following proposition can be useful in showing that some systems are expansive.

Proposition 2.4. — *Let Γ act on a compact metric space $(\Omega, |\,|)$. Suppose that $(U_i)_{i \in I}$ is an open covering of Ω such that for every $i \in I$ there exists some $\gamma_i \in \Gamma$ and $\lambda_i > 1$ satisfying:*

$$|\, \gamma_i x - \gamma_i y \,| \geq \lambda_i |\, x - y \,| \quad \forall x, y \in U_i.$$

Then the system (Ω, Γ) is expansive.

PROOF. By compactness, we can assume I to be finite. We can find an $\epsilon > 0$ (a Lebesgue number of the covering) such that, for every point x of Ω, the open ball centered at x and of radius ϵ is contained in some U_i. Such an ϵ is an expansivity constant for (Ω, Γ). To see this, note that if x and y are two distinct points of Ω which are less than ϵ apart, then x and y are in some U_i and we have

$$|\, \gamma_i x - \gamma_i y \,| \geq \lambda_i |\, x - y \,|.$$

If the distance $|\, \gamma_i x - \gamma_i y \,|$ is again $< \epsilon$, then we can apply the same reasoning to the points $\gamma_i x$ and $\gamma_i y$. Suppose that we can iterate this construction for n steps. Then, we would have $\gamma \in \Gamma$ such that

$$|\, \gamma x - \gamma y \,| \geq \lambda^n |\, x - y \,|,$$

where $\lambda = min\ \lambda_i$. As $\lambda > 1$, there is an integer n for which this blowing up process must stop. Therefore, we have $|\, \gamma x - \gamma y \,| \geq \epsilon$.   ∎

Corollary 2.5. — *Let M be a closed Riemannian manifold (with or without boundary) and let Γ be a semigroup of C^1 maps of M. Assume that for every $x \in M$ there is a $\gamma \in \Gamma$ such that $\|d\gamma(x)\| > 1$. Then the system (Ω, Γ) is expansive.* ∎

The next statement shows that any expansive system admits a "coding" by a subshift.

Proposition 2.6. — *Suppose that the system (Ω, Γ) is expansive. Then there exists a finite set S, a subshift $\Phi \subset \Sigma(\Gamma, S)$ and a continuous, surjective and Γ-equivariant map $\pi : \Phi \to \Omega$.*

PROOF. Let $|\ |$ be a metric on Ω which is compatible with the topology, and $\epsilon > 0$ an expansivity constant for $(\Omega, |\ |, \Gamma)$. By compactness of Ω, we can find a finite covering $(B_s)_{s \in S}$ of Ω with closed sets of diameter $< \epsilon$. Let us denote by $\Sigma(\Gamma, S)$ the set of all maps $\sigma : \Gamma \to S$ and define, for every $\sigma \in \Sigma(\Gamma, S)$,

$$I(\sigma) = \cap_{\gamma \in \Gamma} \gamma^{-1} B_{\sigma(\gamma)}.$$

Let $\Phi \subset \Sigma(\Gamma, S)$ be the set of maps $\sigma : \Gamma \to S$ such that $I(\sigma)$ is nonempty. For every $\sigma \in \Phi$, the set $I(\sigma)$ is reduced to a point. Indeed, if x and y belong to $I(\sigma)$, then γx and γy are in $B_{\sigma(\gamma)}$, and therefore $|\gamma x - \gamma y| < \epsilon$ for every $\gamma \in \Gamma$. As ϵ is an expansivity constant for $(\Omega, |\ |, \Gamma)$, we have $x = y$.

Let $\pi : \Phi \to \Omega$ be the map which to every $\sigma \in \Phi$ associates the unique element in $I(\sigma)$. For every $\gamma \in \Gamma$ and for every $\sigma : \Gamma \to S$, it is clear that $I(\gamma\sigma) = \gamma I(\sigma)$. Consequently, Φ is a Γ-invariant subset of $\Sigma(\Gamma, S)$ and the map π is Γ-equivariant.

Let now (σ_n) be a sequence of elements in Φ which converges to $\sigma \in \Sigma(\Gamma, S)$. For every $\gamma \in \Gamma$, we can find an integer $n(\gamma)$ such that $\sigma_n(\gamma) = \sigma(\gamma)$ for every $n \geq n(\gamma)$. By extracting a subsequence, we can suppose that the sequence $x_n = \pi(\sigma_n)$ converges to a point $x \in \Omega$. For every $\gamma \in \Gamma$, we have $x_n \in \gamma^{-1} B_{\sigma(\gamma)}$ for $n \geq n(\gamma)$. Therefore, $x \in \gamma^{-1} B_{\sigma(\gamma)}$. Hence $x \in I(\sigma)$, which shows that Φ is closed in $\Sigma(\Gamma, S)$ and that the map π is continuous. Finally, the surjectivity of π is a consequence of the fact that the sets (B_s) form a covering of Ω. Indeed, given the elements $x \in \Omega$ and $\sigma : \Gamma \to S$ such that $\gamma x \in B_{\sigma(\gamma)}$ for every $\gamma \in \Gamma$, the definition of π shows that $x = \pi(\sigma)$. ∎

Definition 2.7. — We say that the system (Ω, Γ) is a *quotient* of the system (Ω', Γ) if there exists a continuous, surjective and Γ-equivariant map $\Omega' \to \Omega$.

With this definition, Proposition 2.6 can be formulated in the following way: Any expansive system is the quotient of some subshift.

Remarks.
1) A quotient of an expansive system is not necessarily expansive. For example, consider the $\mathbf{N}$-action on the circle S^1 in the complex plane $\{z \in C \mid |z| = 1\}$ generated by the map $z \mapsto z^2$, and consider also the $\mathbf{N}$-action on the interval $[-1, 1]$ generated by the polynomial $x \mapsto 2x^2 - 1$. By using the vertical projection, the dynamical system $([-1, 1], \mathbf{N})$ appears as a quotient of the system $(S^1, \mathbf{N})$. As the differential of $z \mapsto z^2$ is of norm 2 on S^1, the system $(S^1, \mathbf{N})$ is expansive by Corollary 2.5. But

the polynomial $P(x) = 2x^2 - 1$ generates a dynamical system on $[-1, 1]$ which is not expansive, as opposite points have the same image by P.

Here is now an example with $\Gamma = \mathbb{Z}$ (*cf.* Parry and Walters, [Wal], p. 175). Let F be the homeomorphism of the torus $T^2 = \mathbb{R}^2/\mathbb{Z}^2$ which is induced by the linear automorphism L of $\mathbb{R}^2$ defined by the matrix $\begin{pmatrix} 2 & 1 \\ 1 & 1 \end{pmatrix}$. The action of $\mathbb{Z}$ on T^2 generated by F is expansive (to see this, one can look at L in the neighborhood of $0 \in \mathbb{R}^2$ equipped with a basis of eigenvectors). Let us now consider the involution I of T^2 defined by the symmetry with respect to the origin. The quotient $S^2 = T^2/I$ is a sphere and the projection $T^2 \to S^2$ is a two-sheeted branched covering. F induces a homeomorphism f of S^2 which generates a dynamical system $(S^2, \mathbb{Z})$, which is the quotient of $(T^2, \mathbb{Z})$. But the system $(S^2, \mathbb{Z})$ is not expansive. Indeed, for every $\epsilon > 0$, we can find distinct points x and y in S^2 such that $\mid f^n(x) - f^n(y) \mid \leq \epsilon$ for all $n \in \mathbb{Z}$ (one can take x and y to be the images of two points in $\mathbb{R}^2$ which are symmetric with respect to one of the two eigenspaces of L).

2) If the system $(\Omega, \mathbb{Z})$ is the quotient of a subshift $\Phi \subset \Sigma = \Sigma(\Gamma, S)$, then the topological entropy $h(\Omega, \mathbb{Z})$ is finite. Indeed, we have

$$h(\Omega, \mathbb{Z}) \leq h(\Phi, \mathbb{Z}) \leq h(\Sigma, \mathbb{Z}) = log(card(S)).$$

For example, this implies that the system $([0,1]^{\mathbb{Z}}, \mathbb{Z})$, where $\mathbb{Z}$ acts as the shift on the Hilbert cube $[0,1]^{\mathbb{Z}}$, being of infinite topological entropy, is not the quotient of a subshift (see [Wal] for the definition and the properties of the topological entropy).

3) The topological space Ω cannot be in itself an obstruction to the fact that the system (Ω, Γ) is the quotient of a subshift. Indeed, any compact metrizable space is the image by a continuous map of a Cantor set (see for example [Kur], p. 214, Theorem 4).

Recall that a topological space is said to have *topological dimension zero* if it has a basis whose elements are open and closed sets, and that a compact set has topological dimension zero if and only if it is totally disconnected (see for example [HW]).

The next statement gives a characterization of the systems which are topologically conjugate to a subshift. (Recall that two systems (Ω_1, Γ) and (Ω_2, Γ) are said to be *topologically conjugate* if there exists a Γ-equivariant homeomorphism $\Omega_1 \to \Omega_2$.)

Proposition 2.8. — *For every system (Ω, Γ), the following two conditions are equivalent:*

(*i*)   *There exists a finite set S and a subshift $\Phi \subset \Sigma(\Gamma, S)$ such that (Ω, Γ) is topologically conjugate to (Φ, Γ).*

(*ii*)   *The space Ω has topological dimension zero (i.e. is totally disconnected) and the system (Ω, Γ) is expansive.*

PROOF. The first implication is immediate. Let us prove the second one.

Suppose that Ω has topological dimension zero and that (Ω, Γ) is expansive. Let $|\ |$ be a metric on Ω which is compatible with the topology, and let ϵ be an expansivity constant for $(\Omega, |\ |, \Gamma)$. As Ω has topological dimension 0, we can cover it by sets which are open and closed, and whose diameter is $< \epsilon$. By compactness of Ω, we can extract a finite covering. Thus, we have a partition of Ω by a finite family $(B_s)_{s \in S}$ of closed sets of diameter $< \epsilon$. By looking again at the construction which was done for the proof of Proposition 2.6, we see that the map which associates to every $x \in \Omega$ the unique $\sigma \in \Sigma(\Gamma, S)$ defined by $\gamma x \in B_{\sigma(\gamma)}$ is a Γ-equivariant homeomorphism from Ω to a subshift $\Phi \subset \Sigma(\Gamma, S)$. ∎

§3 – Subshifts of finite type

Following the terminology which is most commonly used in a cartesian product, we shall say that a subset C of Σ is a *cylinder* if there exists a finite subset F of Γ and a set A of maps from F into S such that

$$C = \{\sigma \in \Sigma \mid \sigma_{|F} \in A\}.$$

Remarks.
1) The complement of a cylinder is also a cylinder.
2) The union (resp. intersection) of two cylinders is again a cylinder.
3) Every cylinder is closed and open in Σ.
4) The set of all cylinders forms a basis for the topology of Σ.
5) The cartesian product of two cylinders is again a cylinder. More precisely, if S_1 and S_2 are finite sets, if $C_1 \subset \Sigma(\Gamma, S_1)$ and $C_2 \subset \Sigma(\Gamma, S_2)$ are cylinders, then $C_1 \times C_2 \subset \Sigma(\Gamma, S_1) \times \Sigma(\Gamma, S_2) = \Sigma(\Gamma, S_1 \times S_2)$ is also a cylinder.

Proposition 3.1. — *Let Φ be a subset of Σ. Then, the following three statements are equivalent:*

(i) There exists a cylinder C such that

$$\Phi = \cap_{\gamma \in \Gamma} \gamma^{-1} C.$$

(ii) There exists a continuous and Γ−equivariant map $T : \Sigma \rightarrow \Sigma$ having Φ as its set of fixed points.

(iii) Φ is closed in Σ and there exists an open set U in Σ such that

$$\Phi = \cap_{\gamma \in \Gamma} \gamma^{-1} U.$$

PROOF.
$(i) \Rightarrow (ii)$: We shall suppose that $card(S) \geq 2$ (otherwise, the result is trivial). Let Φ be a map satisfying (i). Then there exists a finite subset F of Γ and a set A of maps from F to S such that

$$\Phi = \{\sigma \in \Sigma \mid \forall \gamma \in \Gamma, \gamma \sigma_{|F} \in A\}.$$

We can suppose that the identity element Id of Γ belongs to F. Let $\mathcal{B}$ be the set of all maps from F to S. It is clear that we can construct a map $u : \mathcal{B} \to S$ satisfying the following property:

(3.1.1) $\forall f \in \mathcal{B}, u(f) = f(Id) \iff f \in A.$

(to define such a map u, we take $u(f) = f(Id)$ if $f \in A$, and $u(f) = $ an element in S which is distinct from $f(Id)$ if $f \notin A$.) Let us now construct the map $T : \Sigma \to \Sigma$. Given $\sigma \in \Sigma$, we define $T(\sigma) : \Gamma \to S$ by

(3.1.2) $\forall \gamma \in \Gamma,\ T(\sigma)(\gamma) = u(\gamma \sigma_{|F}).$

Let us verify that T possesses the required properties. If $\sigma \in \Phi$, we have $\gamma \sigma_{|F} \in A$ for all $\gamma \in \Gamma$, and then, using (3.1.1) and (3.1.2),

$$T(\sigma)(\gamma) = \gamma \sigma(Id) = \sigma(\gamma).$$

Hence $T(\sigma) = \sigma$. Suppose now that $\sigma \notin \Phi$. Then there exists $\gamma \in \Gamma$ such that $\gamma \sigma_F \notin A$. We then have, using (3.1.1) and (3.1.2),

$$T(\sigma)(\gamma) = u(\gamma \sigma_{|F}) \neq \gamma \sigma(Id) = \sigma(\gamma),$$

Hence, $T(\sigma) \neq \sigma$. It follows that Φ is the set of fixed points of T.

Let us show now that T is Γ-equivariant. Let $\eta \in \Gamma$. We have

$$\forall \gamma \in \Gamma,\ \eta T(\sigma)(\gamma) = T(\sigma)(\gamma \eta)$$

$$= u(\gamma \eta \sigma_{|F}) = T(\eta \sigma)(\gamma),$$

which implies that $\eta T(\sigma) = T(\eta \sigma)$. If (σ_n) is a sequence of elements of Σ which converges to $\sigma \in \Sigma$, then, given $\gamma \in \Gamma$, we have, for all n large enough, $\gamma \sigma_{n|F} = \gamma \sigma_{|F}$, and therefore, using (3.1.2), $T(\sigma_n)(\gamma) = T(\sigma)(\gamma)$. This proves the continuity of T.

$(ii) \Rightarrow (iii)$: Let $T : \Sigma \to \Sigma$ be a continuous Γ-equivariant map and let $\Phi = Fix(T)$ (the set of fixed points of T). Define also, for every $s \in S$,

$$U_s = \{\sigma \in \Sigma \mid \sigma(Id) = s\}\ and\ V_s = T^{-1}(U_s).$$

We have $\sigma \in \Phi$ if and only if, for every $\gamma \in \Gamma$, $T(\sigma)(\gamma) = \sigma(\gamma)$. On the other hand, we have $T(\sigma)(\gamma) = \gamma T(\sigma)(Id) = T(\gamma \sigma)(Id)$ and $\sigma(\gamma) = \gamma \sigma(Id)$. Therefore:

(3.1.3) $\Phi = \cap_{\gamma \in \Gamma} \gamma^{-1} U,$

where $U = \bigcup_{s \in S} V_s \cap U_s$. As U_s is a cylinder, the sets U_s and V_s are open and closed in Σ. Equality (3.1.3) implies then that Φ satisfies (iii).

$(iii) \Rightarrow (i)$: Let Φ be a closed set and U an open set in Σ such that $\Phi = \cap_{\gamma \in \Gamma} \gamma^{-1} U$. The set U is the union of the cylinders that it contains. By compactness of Φ, there exists a finite family of cylinders $(C_i)_{i \in I}$ contained in U which cover the set Φ. Let C be the union of the C_i. (Note that C is therefore a cylinder.) We have $\Phi = \cap_{\gamma \in \Gamma} \gamma^{-1} C$, which implies (i). ∎

Exercise. Show directly that $(ii) \Rightarrow (i)$, by using Proposition 1.4.

Let us note that any subset Φ of Σ satisfying any of the equivalent conditions of the preceding proposition is a subshift, as Φ is closed in Σ and is Γ−invariant.

Definition 3.2. — We say that the subset Φ of Σ is a *subshift of finite type* if it satisfies the equivalent conditions of Proposition 3.1.

Remark. A subshift of finite type is called also a *Markovian* subshift.

Exercise. Show that every subshift is the intersection of a countable family of subshifts of finite type.

Exercise. Show that the intersection of two subshifts of finite type is also a subshift of finite type.

Exercise. Let $\Gamma = \mathbb{Z}$ and $S = \{0,1,2\}$; we have therefore $\Sigma = \{0,1,2\}^{\mathbb{Z}}$. Let $\Phi_1 = \{0,1\}^{\mathbb{Z}}$ and $\Phi_2 = \{1,2\}^{\mathbb{Z}}$. Show that Φ_1 and Φ_2 are subshifts of finite type, but that $\Phi_1 \cup \Phi_2$ is not a subshift of finite type.

Let $\Phi \subset \Sigma(\Gamma, S)$. Then, the diagonal $\Delta(\Phi)$ of $\Phi \times \Phi$, defined as

$$\Delta(\Phi) = \{(\sigma, \sigma) \mid \sigma \in \Phi\},$$

is a subset of

$$\Sigma(\Gamma, S) \times \Sigma(\Gamma, S) = \Sigma(\Gamma, S \times S).$$

Proposition 3.3. — *If Φ is subshift of finite type, then the same is true for the diagonal $\Delta(\Phi)$ of $\Phi \times \Phi$.*

PROOF. Let $T : \Sigma \to \Sigma$ be a continuous Γ−equivariant map such that $\Phi = Fix(T)$. It is sufficient to remark that $\Delta(\Phi) = Fix(T')$, where $T' : \Sigma \times \Sigma \to \Sigma \times \Sigma$ is defined by the formula $T'((\sigma, \sigma')) = (\sigma', T(\sigma))$. ∎

Consider now two finite sets S_1 and S_2, and two subsets $\Phi_1 \subset \Sigma(\Gamma, S_1)$ and $\Phi_2 \subset \Sigma(\Gamma, S_2)$. Then, $\Phi_1 \times \Phi_2$ is a subset of $\Sigma(\Gamma, S_1) \times \Sigma(\Gamma, S_2) = \Sigma(\Gamma, S_1 \times S_2)$.

Proposition 3.4. — *If Φ_1 and Φ_2 are subshifts of finite type, then the same is true for $\Phi_1 \times \Phi_2$.*

PROOF. For $i = 1, 2$, let T_i be a continuous and Γ-equivariant map from $\Sigma(\Gamma, S_i)$ to itself, such that $\Phi_i = Fix(T_i)$. The proposition follows from the fact that $\Phi_1 \times \Phi_2 = Fix(T_1 \times T_2)$. ∎

§4 – Systems of finite type and finitely presented systems

Let again Ω be a compact metrizable space on which Γ acts continuously.

Definition 4.1. — We say that (Ω, Γ) is a *system of finite type* if there exists a finite set S, a subshift of finite type $\Phi \subset \Sigma(\Gamma, S)$ and a continuous, surjective and Γ-equivariant map $\pi : \Phi \to \Omega$.

Examples.
1) If Φ is a subshift of finite type, then (Φ, Γ) is a system of finite type (take, as a map π, the identity map).
2) Given the systems (Ω, Γ) and (Ω', Γ), suppose there exists a continuous, surjective and Γ-equivariant map $f : \Omega \to \Omega'$ (*i.e.* suppose that (Ω', Γ) is a quotient of (Ω, Γ)) and that (Ω, Γ) is of finite type. Then, (Ω', Γ) is of finite type.
3) If the systems (Ω, Γ) and (Ω', Γ) are of finite type, then the same is true for the system $(\Omega \times \Omega', \Gamma)$ (use Proposition 3.4).

Consider two systems (Φ, Γ) and (Ω, Γ) and a map $\pi : \Phi \to \Omega$ which is continuous, surjective and Γ-equivariant. Let us denote by $R(\pi)$ the (graph of the) equivalence relation associated to π,

$$R(\pi) = \{(\varphi_1, \varphi_2) \in \Phi \times \Phi \mid \pi(\varphi_1) = \pi(\varphi_2)\}.$$

This is a closed and Γ-invariant subset of $\Phi \times \Phi$, and π induces a Γ-equivariant homeomorphism from the quotient space $\Phi / R(\pi)$ to Ω.

In the case where Φ is a subshift of $\Sigma(\Gamma, S)$, we note that $R(\pi)$ is a subshift of $\Sigma(\Gamma, S) \times \Sigma(\Gamma, S) = \Sigma(\Gamma, S \times S)$

Definition 4.2. — We say that (Ω, Γ) is a *finitely presented system* if there exists a finite set S, a subshift of finite type $\Phi \subset \Sigma(\Gamma, S)$ and a continuous, surjective and Γ-equivariant map $\pi : \Phi \to \Omega$ such that the subshift $R(\pi)$ is of finite type. Such a map $\pi : \Phi \to \Omega$ is called a *finite presentation* of the system (Ω, Γ).

Examples.
1) If Φ is a subshift of finite type, then (Φ, Γ) is a finitely presented system (take, as a map π, the identity map and use Proposition 3.3).
2) If the systems (Ω, Γ) and (Ω', Γ) are finitely presented, then the same is true for the system $(\Omega \times \Omega', \Gamma)$ (use Proposition 3.4).
3) Let $\Omega \subset V$ be a basic set of a diffeomorphism f of a compact manifold V, satisfying Smale's Axiom A (*cf.* [Sma]). Then the restriction of f to Ω generates a finitely presented system $(\Omega, \mathbb{Z})$. This can be shown using a partition into Markov rectangles for Ω (see [Bow] and [Fri]). In the same way, using Markov partitions for pseudo-Anosov homeomorphisms of surfaces, one can show that these maps generate finitely presented $\mathbb{Z}$-systems.

Proposition 4.3. — *Let $\Phi \subset \Sigma(\Gamma, S)$ be a subshift of finite type and $\pi : \Phi \to \Omega$ a continuous, surjective and Γ-equivariant map. Then the subshift $R(\pi)$ is of finite type if and only if the system (Ω, Γ) is expansive.*

In the proof, we shall use the following

Lemma 4.4. — *Let $\Phi \subset \Sigma(\Gamma, S)$ be a subshift of finite type and Φ' a closed subset of Φ. Then, Φ' is a subshift of finite type if and only if there exists an open set U in Φ such that $\Phi' = \cap_{\gamma \in \Gamma} \gamma^{-1} U$.*

PROOF. The condition is necessary. Indeed, if Φ' is a subshift of finite type, there exists, by definition, an open subset V of $\Sigma(\Gamma, S)$ such that $\Phi' = \cap_{\gamma \in \Gamma} \gamma^{-1} V$. The set $U = V \cap \Phi$ is open in Φ and we have $\Phi' \subset \cap_{\gamma \in \Gamma} \gamma^{-1} U \subset \cap_{\gamma \in \Gamma} \gamma^{-1} V$, hence $\Phi' = \cap_{\gamma \in \Gamma} \gamma^{-1} U$.

Let us show now that the condition is sufficient. Let U be an open subset of Φ such that $\Phi' = \cap_{\gamma \in \Gamma} \gamma^{-1} U$ and let V be an open subset of $\Sigma(\Gamma, S)$ such that $U = V \cap \Phi$. As Φ is a subshift of finite type, there exists an open subset V_0 of $\Sigma(\Gamma, S)$ such that $\Phi = \cap_{\gamma \in \Gamma} \gamma^{-1} V_0$. We have

$$\Phi' = \cap_{\gamma \in \Gamma} \gamma^{-1}(V \cap \Phi)$$

$$= (\cap_{\gamma \in \Gamma} \gamma^{-1} V) \cap \Phi$$

$$= (\cap_{\gamma \in \Gamma} \gamma^{-1} V) \cap (\cap_{\gamma \in \Gamma} \gamma^{-1} V_0)$$

$$= \cap_{\gamma \in \Gamma} \gamma^{-1}(V \cap V_0),$$

which shows that Φ' is a subshift of finite type. ∎

Proof of Proposition 4.3 Let us first remark that it follows from Proposition 3.4 that $\Phi \times \Phi$ is a subshift of finite type. Let $P = \pi \times \pi : \Phi \times \Phi \to \Omega \times \Omega$ and let Δ be the diagonal of $\Omega \times \Omega$. We have $R(\pi) = P^{-1}(\Delta)$.

Suppose that $R(\pi)$ is of finite type. Then, there exists, according to the preceding lemma, an open subset U of $\Phi \times \Phi$ such that $R(\pi) = \cap_{\gamma \in \Gamma} \gamma^{-1} U$. Let $K = P(\Phi \times \Phi - U)$. K is compact and has empty intersection with Δ. Therefore, $V = \Omega \times \Omega - K$ is an open subset of $\Omega \times \Omega$ which contains Δ. We have

$$R(\pi) = \cap_{\gamma \in \Gamma} \gamma^{-1} P^{-1}(V) \text{ (because } R(\pi) \subset P^{-1}(V) \subset U)$$

$$= \cap_{\gamma \in \Gamma} \gamma P^{-1}(\gamma^{-1} V), \text{ (because } P \text{ is } \Gamma - \text{equivariant)}$$

$$= P^{-1}(\cap_{\gamma \in \Gamma} \gamma^{-1} V).$$

Therefore, $\Delta = P(R(\pi)) = \cap_{\gamma \in \Gamma} \gamma^{-1} V$, which shows that (Ω, Γ) is expansive.

Let us show now that the condition is sufficient. If (Ω, Γ) is expansive, there exists, by definition, an open subset V of $\Omega \times \Omega$ such that $\Delta = \cap_{\gamma \in \Gamma} \gamma^{-1} V$. Let $U = P^{-1}(V)$. We have

$$R(\pi) = P^{-1}(\Delta)$$

$$= P^{-1}(\cap_{\gamma \in \Gamma} \gamma^{-1} V)$$

$$= \cap_{\gamma \in \Gamma} P^{-1}(\gamma^{-1}V)$$
$$= \cap_{\gamma \in \Gamma} \gamma^{-1} P^{-1}(V)$$
$$= \cap_{\gamma \in \Gamma} \gamma^{-1} U,$$

which shows that $R(\pi)$ is of finite type, using the fact that U is an open subset of $\Phi \times \Phi$ (Lemma 4.4). ∎

Proposition 4.3 gives immediately the following corollaries.

Corollary 4.5. — *the system (Ω, Γ) is finitely presented if and only if it is both expansive and of finite type.* ∎

Remark. Let F be the homeomorphism of the torus $T^2 = \mathbb{R}^2/\mathbb{Z}^2$ whose lift to $\mathbb{R}^2$ is the linear automorphism whose matrix is $\begin{pmatrix} 2 & 1 \\ 1 & 1 \end{pmatrix}$. Let f be the homeomorphism of the sphere $S^2 = T^2/\pm Id$ induced by F. The system $(T^2, \mathbb{Z})$ which is generated by F is finitely presented (use the fact that F satisfies Smale's Axiom A with $\Omega = T^2$ as a basic set, *cf.* [Bow]). But the system $(S^2, \mathbb{Z})$ generated by f is not finitely presented (because as we saw above, it is not expansive), although it is of finite type (being the quotient of a system of finite type). A quotient of a finitely presented system is therefore not necessarily finitely presented.

Corollary 4.6. — *Let (Ω, Γ) be a finitely presented system. If Φ is a subshift of finite type and if $\pi : \Phi \to \Omega$ is a continuous, surjective and Γ-equivariant map, then the subshift $R(\pi)$ is of finite type.* ∎

Let us note finally that for fixed Γ and S, the set of cylinders, and therefore the set of subshifts of finite type of $\Sigma(\Gamma, S)$ is countable. We therefore have the following

Proposition 4.7. — *If the semigroup Γ is fixed, then the class of finitely presented dynamical systems (Ω, Γ) up to topological conjugacy is countable.*

§5 – Symbolic dynamics on N and on Z

In all this section, Γ is either the semigroup $\mathbf{N}$ of nonnegative integers, or the group $\mathbb{Z}$ of all integers. We fix a finite set S. For $\Gamma = \mathbf{N}$ (resp. $\mathbb{Z}$), the shift $\Sigma = \Sigma(\Gamma, S)$ is called the *one-sided* (resp. *two-sided*) Bernoulli shift. Let us recall also that Σ is the set of sequences $\sigma : \Gamma \to S$ and that the action of Γ on Σ is given by

$$\gamma\sigma(i) = \sigma(i + \gamma) \text{ with } i, \gamma \in \Gamma.$$

Recall also that a *word* of length n on S ($n \geq 0$) is, by definition, an element of S^n, that is, a sequence $w = (s_1, ..., s_n)$ of n elements in S. We shall say that the word w *appears* in $\sigma \in \Sigma$ if there exists an integer $i \in \Gamma$ such that $(\sigma(i), ...\sigma(i+n-1)) = w$. Let $S^* = \cup_{n \geq 0} S^n$ be the set of all words on S.

Proposition 5.1. — *A subset Φ of Σ is a subshift if and only if there exists a set of words $W \subset S^*$ such that Φ is the set of $\sigma \in \Sigma$ in which no word of W appears.*

PROOF. For every set of words $W \subset S^*$, let us define

$$\Phi(W) = \{\sigma \in \Sigma \mid \forall w \in W, w \text{ does not appear in } \sigma\}.$$

It is clear that $\Phi(W)$ is a closed Γ-invariant subset of Σ, and is therefore a subshift.

Conversely, let Φ be a subshift. Let W be the set of words which do not appear in any element of Φ. It is clear that $\Phi \subset \Phi(W)$. Consider now an element $\sigma \in \Phi(W)$. For every integer $n \geq 0$, we can find an element of Φ in which the word $(\sigma(0), \sigma(1), ..., \sigma(n))$ appears (resp., if $\Gamma = \mathbb{Z}$, the word $(\sigma(-n), \sigma(-n+1), ..., \sigma(n))$). As Φ is Γ-invariant, we deduce that there exists an element $\sigma_n \in \Phi$ such that $\sigma_n(i) = \sigma(i)$ for every i such that $\mid i \mid \leq n$. The sequence (σ_n) converges to σ. It follows from the fact that Φ is closed that $\sigma \in \Phi$, and this finishes the proof. ∎

Given a set $W \subset S^n$ of words of length n, we define

$$\Sigma_W = \{\sigma \in \Sigma \mid \forall i \in \Gamma, (\sigma(i), ..., \sigma(i+n-1)) \in W\}.$$

Proposition 5.2. — *Let Φ be a subset of Σ. The following three statements are equivalent:*

(i) Φ is a subshift of finite type.

(ii) There exists an integer n and a subset $W \subset S^n$ such that $\Phi = \Sigma_W$.

(iii) There exists an integer n and a subset $W' \subset S^n$ such that Φ is the set of $\sigma \in \Sigma$ in which no word of W' appears.

PROOF. The equivalence of (ii) and (iii) can be seen by taking $W' = S^n - W$. Given $W \subset S^n$, let C be the cylinder defined as the set of elements $\sigma \in \Sigma$ such that $(\sigma(0), \sigma(1), ..., \sigma(n-1)) \in W$. We have $\Sigma_W = \cap_{\gamma \in \Gamma} \gamma^{-1} C$, which shows that Σ_W is a subshift of finite type and therefore that (ii) implies (i).

Conversely, if Φ is a subshift of finite type, there exists a finite set $F \subset \Gamma$ and a set A of maps from F to S such that Φ is the set of $\sigma \in \Sigma$ such that the restriction of $\gamma\sigma$ to F belongs to A for every $\gamma \in \Gamma$. It is clear that we can suppose that $F = \{0, 1, 2, ..., n-1\}$ for a certain n. We remark then that $\Phi = \Sigma_W$, where W is the set of words of the form $(a(0), a(1), ..., a(n-1)), a \in A$, which finishes the proof. ∎

Example. Let us take $S = \{0, 1\}$ and $\Gamma = \mathbb{Z}$. Let Φ be the set of sequences $\mathbb{Z} \mapsto S$ having at most one zero term. It is clear that Φ is a subshift. But Φ is not a subshift of finite type. Indeed, suppose that $\Phi = \Sigma_W$ for $W \subset S^n$. Let $\sigma : \mathbb{Z} \to S$ be the sequence whose terms are all zero except $\sigma(0)$ and $\sigma(n)$, which are equal to 1. We have $\sigma \notin \Phi$, and nevertheless every subword of length n which appears in σ appears also in an element of Φ, which is a contradiction.

Note that the system $(\Phi, \mathbb{Z})$ is topologically conjugate to $(\mathbb{Z} \cup \{\infty\}, \mathbb{Z})$, where the action of $\mathbb{Z}$ on $\mathbb{Z} \cup \{\infty\}$ is generated by the translation $z \to z + 1$.

Definition 5.3. — We say that a subshift $\Phi \subset \Sigma$ is of order n if there exists a subset $W \subset S^n$ such that $\Phi = \Sigma_W$.

The following proposition allows us to reduce the study of subshifts of finite type to that of subshifts of order two.

Proposition 5.4. — *Let $\Phi \subset \Sigma(\Gamma, S)$ be a subshift of finite type. Then, there exists a finite set S' and a subshift of order two $\Phi' \subset \Sigma(\Gamma, S')$ such that (Φ, Γ) is topologically conjugate to (Φ', Γ).*

PROOF. Suppose that Φ is of order n. Let $W \subset S^n$ be such that $\Phi = \Sigma_W$. We take as a new set of symbols $S' = W$, and we define $\Sigma' = \Sigma(\Gamma, S')$. Consider the map $u : \Phi \to \Sigma'$ which associates to every $\sigma \in \Phi$ the sequence $u(\sigma) : \Gamma \to S'$ defined by:

$$u(\sigma)(i) = \big(\sigma(i), \sigma(i+1), ..., \sigma(i+n-1)\big), \text{ for every } i \in \Gamma.$$

Let $W' \subset S'^2$ be the set of words of length two on S' which are of the form

$$\big(u(\sigma)(0), u(\sigma)(1)\big)$$

with σ describing any element of Φ. It is clear that u is a Γ-equivariant homeomorphism from Φ to the order-two subshift $\Phi' = \Sigma'_{W'}$. ∎

A subshift of order two can be described by a square matrix with coefficients in $\{0, 1\}$. To see this, we begin by noting that a subset $W \subset S^2$ of words of length two can be defined by its *characteristic function* $M : S^2 \to \{0, 1\}$ given by

$$M(s, s') = 1 \text{ if } (s, s') \in W,$$

$$M(s, s') = 0 \text{ if not.}$$

One can see M as a square matrix having S as a set of indices for its rows and columns, and with coefficients in $\{0, 1\}$. Let us denote by Σ_M the subshift of order two $\Phi = \Sigma_M$ which is thus associated to the matrix M. We have therefore

$$\Sigma_M = \{\sigma \in \Sigma \mid \forall i \in \Gamma, M(\sigma(i), \sigma(i+1)) = 1\}.$$

We say that M is a *transition matrix* for $\Phi = \Sigma_M$.

Examples.
In all the examples which follow, we take $S = \{1, 2\}$ and $\Gamma = \mathbb{Z}$.

1) For $M = \begin{pmatrix} 0 & 0 \\ 0 & 0 \end{pmatrix}$, $\begin{pmatrix} 0 & 1 \\ 0 & 0 \end{pmatrix}$ or $\begin{pmatrix} 0 & 0 \\ 1 & 0 \end{pmatrix}$, Σ_M is empty.

2) For $M = \begin{pmatrix} 1 & 0 \\ 0 & 0 \end{pmatrix}$, $\begin{pmatrix} 1 & 1 \\ 0 & 0 \end{pmatrix}$, $\begin{pmatrix} 1 & 0 \\ 1 & 0 \end{pmatrix}$, $\begin{pmatrix} 0 & 0 \\ 1 & 1 \end{pmatrix}$, $\begin{pmatrix} 0 & 1 \\ 0 & 1 \end{pmatrix}$ or $\begin{pmatrix} 0 & 0 \\ 0 & 1 \end{pmatrix}$, Σ_M is rduced to a point.

3) Forr $M = \begin{pmatrix} 1 & 0 \\ 0 & 1 \end{pmatrix}$, Σ_M is reduced to two points (which are fixed by the action of $\mathbb{Z}$).

4) For $M = \begin{pmatrix} 0 & 1 \\ 1 & 0 \end{pmatrix}$, Σ_M is, as in the preceding example, reduced to two points, but the action of $\mathbb{Z}$ interchanges the two points.

5) For $M = \begin{pmatrix} 1 & 1 \\ 0 & 1 \end{pmatrix}$ or $\begin{pmatrix} 1 & 0 \\ 1 & 1 \end{pmatrix}$, Σ_M presents a $North - South$ dynamics: $(\Sigma_M, \mathbb{Z})$ is topologically conjugate to $(\mathbb{Z} \cup \{-\infty, +\infty\}, \mathbb{Z})$.

6) For $M = \begin{pmatrix} 1 & 1 \\ 1 & 1 \end{pmatrix}$, $\Sigma_M = \Sigma$.

7) For $M = \begin{pmatrix} 0 & 1 \\ 1 & 1 \end{pmatrix}$ or $\begin{pmatrix} 1 & 1 \\ 1 & 0 \end{pmatrix}$, the two systems $(\Sigma_M, \mathbb{Z})$ which are obtained are topologically conjugate (this is done by interchanging the two elements of S). Σ_M is a Cantor set. The dynamics of the action is chaotic. The same as on Σ, the union of the periodic orbits is dense in Σ_M. Note that $(\Sigma_M, \mathbb{Z})$ is not topologically conjugate to $(\Sigma, \mathbb{Z})$ (Σ has two fixed points, whereas Σ_M has only one).

The set $W \subset S^2$ can also be seen as an oriented graph whose set of vertices is S and whose edges are the pairs of elements $(s, s') \in W$. We say that this graph is a *transition graph* for the order-two subshift Σ_W.

Examples. For $S = \{1, 2, 3\}$ and $M = \begin{pmatrix} 1 & 1 & 0 \\ 0 & 0 & 1 \\ 1 & 1 & 0 \end{pmatrix}$, the associated graph is given in Figure 1.

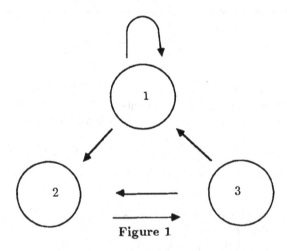

Figure 1

Let M be a square matrix having S as a set of indices for its rows and columns, and whose coefficients are in $\{0, 1\}$. We can easily "read" on M certain dynamical properties of the order-two subshift Σ_M. As an example, let us show how to calculate the ζ-function of Σ_M.

For the rest of this section, we shall restrict to $\mathbb{Z}$-systems.

Recall (see [Sma]) that the ζ-*function* of a $\mathbb{Z}$-system $(\Omega, \mathbb{Z})$ is the formal power-series

$$\zeta(t) = exp(\Sigma_{n \geq 1} n^{-1} N_n t^n)$$

where N_n is the number of elements in Ω whose period is n (*i.e.* the fixed points under the action of the group $n\mathbb{Z} \subset \mathbb{Z}$ of multiples of n). We note that the ζ-function is related to the formal power-series

$$N(t) = \Sigma_{n \geq 1} N_n t^n$$

by the formula

$$N(t) = t\zeta'(t)/\zeta(t).$$

Proposition 5.5. — *The ζ-function of the subshift of order two $(\Sigma_M, \mathbb{Z})$ is given by the formula*

$$\zeta(t) = det(I - tM)^{-1}$$

where $I : S \times S \to \{0,1\}$ is the identity matrix, that is, the matrix defined by $I(s, s') = 1$ if and only if $s = s'$.

To prove this proposition, we use the following lemma which follows immediately from the definition of the product of two matrices.

Lemma 5.6. — *Let $s, s' \in S$ and n an integer which is ≥ 1. The number of words $w = (s_1, ..., s_n) \in S^n$ such that $M(s_i, s_{i+1}) = 1$, for every $i = 1, 2, ..., n-1$, $s_1 = s$ and $s_n = s'$ is equal to $M^n(s, s')$.* ∎

Proof of Proposition 5.5. From the preceding lemma, we know that the number of elements $\sigma \in \Sigma_M$ which are fixed under the action of $n\mathbb{Z}$ and such that $\sigma(0) = s$ is equal to $M^n(s, s)$. We therefore have $N_n = Tr\ M^n$. This gives

$$\zeta(t) = exp(\Sigma_{n \geq 1} n^{-1}(TrM^n)t^n)$$

$$= exp(Tr(\Sigma_{n \geq 1} n^{-1} t^n M^n))$$

$$= det(I - tM)^{-1}.$$

∎

From the fact that ζ is invariant under topological conjugacy, the preceding proposition, together with Proposition 5.4, gives the following

Corollary 5.7. — *The ζ-function of every subshift of finite type on $\mathbb{Z}$ is of the form*

$$\zeta(t) = det(I - tM)^{-1}$$

where M is a square matrix with coefficients in $\{0,1\}$. ∎

Corollary 5.8. — *The ζ-function of every subshift of finite type on $\mathbb{Z}$ is a rational function.*

The preceding corollary is a particular case of the celebrated rationality theorem of Manning, which can be stated in the following way:

Theorem 5.9. — *Let $(\Omega, \mathbb{Z})$ be a finitely presented $\mathbb{Z}$-system, and suppose that there is a finite presentation $\pi : \Phi \to \Omega$ which is finite-to-one, that is, satisfies the following condition (F):*

(F) *There exists an integer N such that*

$$card\big(\pi^{-1}(x)\big) \leq N \text{ for all } x \in \Omega.$$

Then, the ζ-function of $(\Omega, \mathbb{Z})$ is rational.   ∎

For a proof of the theorem of Manning, the reader is referred to [Man] and to [Fri] (see also section 8.5.U of [Gro 3]).

In fact, D. Fried has proved in [Fri] that any finitely presented system $(\Omega, \mathbb{Z})$ admits a finite presentation $\pi : \Phi \mapsto \Omega$ which satisfies (F). Hence, any finitely presented $\mathbb{Z}$-system has a rational ζ-function.

§6 – Sofic systems

In this section, Γ is again an arbitrary countable semigroup. Let S be a finite set and let $\Phi \subset \Sigma(\Gamma, S)$ be a subshift. We say that Φ is a *sofic* subshift if the dynamical system (Φ, Γ) is a finitely presented system (the terminology *sofic* is due to B. Weiss [Wei], and comes from the hebrew word $SOFI$ which means "finite" or "having an end"). We note that the subshift Φ is sofic if and only if there exists a finite set S_0, a subshift of finite type $\Phi_0 \subset \Sigma(\Gamma, S_0)$ and a continuous, surjective and Γ-equivariant map $\pi : \Phi_0 \to \Phi$. Indeed, Proposition 4.3 shows that such a map π is a finite presentation of (Φ, Γ).

Example. Let us look again at the subshift which we gave as an example in the preceding section. We take $\Gamma = \mathbb{Z}$, $S = \{0,1\}$ and we let Φ be the subshift $\subset \Sigma = \Sigma(\Gamma, S)$ which is defined as the set of sequences $\sigma \in \Sigma$ which have at most one nonzero term. The system $(\Phi, \mathbb{Z})$ is topologically conjugate to $(\mathbb{Z} \cup \{\infty\}, \mathbb{Z})$.

Consider now the subshift of order two $\Phi_0 \subset \Sigma$ which consists of the sequences $s \in \Sigma$ such that $\big(\sigma(i), \sigma(i+1)\big) \neq (1,0)$ for every $i \in \mathbb{Z}$. The system $(\Phi_0, \mathbb{Z})$ is topologically conjugate to $(\mathbb{Z} \cup \{-\infty, \infty\}, \mathbb{Z})$. The map

$$\pi : \Phi_0 = \mathbb{Z} \cup \{-\infty, \infty\} \to \mathbb{Z} \cup \{\infty\} = \Phi$$

defined by $\pi(-\infty) = \pi(\infty) = \infty$ and $\pi(z) = z$ for all $z \in \mathbb{Z}$, is a finite presentation of $(\Phi, \mathbb{Z})$. As a consequence, Φ is a sofic subshift. The map π can be read on the labelled graph of Figure 2, in the following way. This graph is a transition graph for Φ_0, equipped with a *labelling* of the edges, with values in S, that is, a map $E \to S$,

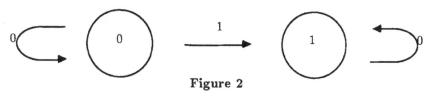

Figure 2

where E stand for the set of edges of the graph. The map $\pi : \Phi_0 \to \Phi$ is obtained by associating to each element of Φ_0, that is, to each $\mathbb{Z}$-path in the graph, the sequence obtained by reading successively the labelling on the edges of that path.

Exercise. We use the notations of the preceding example.
1) Explicit the relation $R(\pi) = \{(\sigma, \sigma') \in \Phi_0 \times \Phi_0 \mid \pi(\sigma) = \pi(\sigma')\}$, and show directly that $R(\pi)$ is a subshift of finite type of $\Sigma \times \Sigma$.
2) Show that Φ is not topologically conjugate to any subshift of finite type, that is, that we cannot find a finite set S' and a subshift of finite type $\Phi' \subset \Sigma(\mathbb{Z}, S')$ such that $(\Phi, \mathbb{Z})$ is topologically conjugate to $(\Phi', \mathbb{Z})$.

The dynamical system (Ω, Γ) is said to be *sofic* if there exists a finite set S and a *sofic* subshift $\Phi \subset \Sigma(\Gamma, S)$ such that (Ω, Γ) is topologically conjugate to (Φ, Γ). We have the following characterization of sofic systems, which is an immediate consequence of Proposition 2.8.

Proposition 6.1. — *For a system (Ω, Γ) to be sofic, it is necessary and sufficient that the space Ω is of topological dimension 0 (i.e. that this space is totally disconnected) and that the system (Ω, Γ) is finitely presented.* ∎

Sofic systems on $\mathbb{Z}$ and on $\mathbf{N}$.

In the rest of this section, we take $\Gamma = \mathbb{Z}$ or $\mathbf{N}$. We shall describe the sofic subshifts with the help of graphs with labelled vertices. A *graph with labelled vertices* (G, μ) is, by definition, an oriented finite graph G, that is, a finite set of vertices V and a set of vertices $E \subset V \times V$, together with a map $\mu : V \to S$ from the set of vertices to a finite set S. To such a labelled graph (G, μ), we associate a sofic subshift $\Phi \subset \Sigma(\Gamma, S)$ and a finite presentation $\pi : \Phi_0 \to \Phi$ in the following manner:

Let $\Phi_0 \subset \Sigma(\Gamma, V)$ be the subshift of order two having G as a transition graph. We associate to every element σ of Φ_0 the sequence $\pi(\sigma) : \Gamma \to S$ defined by $\pi(\sigma)(i) = \mu(\sigma(i))$ for $i \in \Gamma$. It is clear that the set Φ of $\pi(\sigma), \sigma \in \Phi_0$, is a subshift of $\Sigma(\Gamma, S)$ and that the map $\pi : \Phi_0 \to \Phi$ is continuous, surjective and Γ-equivariant. The sofic subshift Φ is called the subshift associated to (G, μ). In fact, we have the following

Proposition 6.2. — *A subshift $\Phi \subset \Sigma(\Gamma, S)$ is sofic if and only if there exists a graph with labelled vertices (G, μ) such that Φ is the subshift associated to (G, μ).*

PROOF. It remains to show that the condition is necessary. Let $\Phi \subset \Sigma(\Gamma, S)$ be a sofic subshift. There exists therefore a finite set S_1, a subshift of finite type $\Phi_1 \subset \Sigma(\Gamma, S_1)$ and a map $\pi : \Phi_1 \to \Phi$ which is continuous, surjective and Γ-equivariant. By Proposition 1.4, there exists a finite set $F \subset \Gamma$ and a map $u : S_1^F \to S_2$ such that $\pi = u_\infty$, *i.e.* such that $\pi(\sigma)(\gamma) = u(\gamma\sigma_{|F})$ for every $\sigma \in \Phi_1$ and for every $\gamma \in \Gamma$. At the expense of taking F larger and of composing π, to the right, by an element of Γ, we can suppose that $F = \{0, 1, ..., n-1\}$ and that the subshift Φ_1 is of order n. The set S_1^F is then identified with the set of words of length n on S_1. Consider now, as a new set of symbols, the set S_0 of words of length n on S_1 which appear in at least one element of Φ_1, and let μ be the restriction of u to S_0. The map which associates to every element σ of Φ_1 the sequence $\tau : \Gamma \to S_0$ defined by $\tau(i) = \big(\sigma(0), \sigma(1), ..., \sigma(n-1)\big)$ is a topological conjugacy between Φ_1 and a subshift of order two $\Phi_0 \subset \Sigma(\Gamma, S_0)$ (proof of Proposition 5.4). If G is a transition graph for Φ_0, it is clear that Φ is the subshift associated to (G, μ). ∎

Example. Let again $\Gamma = \mathbb{Z}$, $S = \{0, 1\}$ and $\Phi \subset \Sigma(\Gamma, S)$ the subshift consisting of the sequences having at most one non-zero term. Φ is the sofic subshift associated to the graph with labelled vertices of Figure 3.

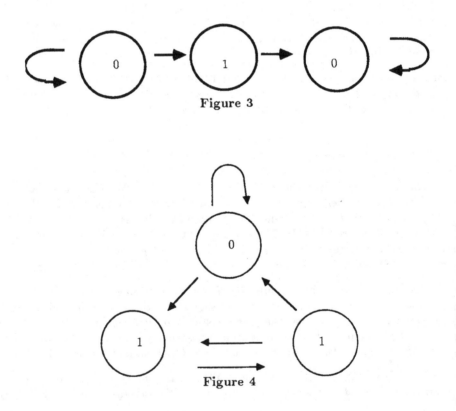

Figure 3

Figure 4

Exercise. We take $\Gamma = \mathbb{Z}$ and $S = \{0,1\}$. Let $\Phi \subset \Sigma(\Gamma, S)$ be the set of sequences $\Gamma \to S$ satisfying the following property: between any two zero terms, there is always an even number of non-zero terms. Prove that Φ is the sofic subshift associated to the graph with labelled vertices given in Figure 4. Calculate the ζ-function of Φ. Deduce that Φ is not topologically conjugate to any subshift of finite type (in case of difficulty, see [CP]).

Notes and comments on Chapter 2

The definitions and the statements on symbolic dynamics which are given in this chapter are, for the most part, well-known, at least for the case where $\Gamma = \mathbb{Z}$ or $\mathbb{N}$ (see for instance [Bro], [DGS], [Wal] and the bibliography there). The extension to the case of an arbitrary countable semigroup Γ is immediate. Proposition 1.4 is due to Hedlund and Lyndon ([Hed], Theorem 3.4). The definition of a subshift of finite type as the set of fixed points of a continuous and equivariant map from the ambient Bernoulli shift to itself (property (ii) of Proposition 3.1) is contained in §5.1 of [Gro 1]. The notion of finitely presented dynamical system appears in [Gro 1] (§5.2, where such a system is called a *hyperbolic system*) and in [Gro 3] (§8.4.B in a more general setting). Proposition 4.3 is Lemma 1 of [Fri]. The rationality of the ζ-function for subshifts of finite type is due to Bowen and Lanford [BL]. The notion of sofic system is due to B. Weiss ([Wei]). See [Fis] for the use of graphs in the study of sofic systems. In [CP], Coven and Paul show that any sofic system on $\mathbb{Z}$ or on $\mathbb{N}$ admits a finite presentation satisfying condition (F) of the rationality theorem of Manning (Theorem 5.9). In [BL], there is a description, for any finite set S whose cardinality is ≥ 2, of an uncountable family of subshifts $\Phi \subset \Sigma(\mathbb{Z}, S)$ whose associated ζ-functions are all irrational and distinct. This proves in particular, using Corollary 4.5, the existence of an uncountable family of expansive $\mathbb{Z}$-systems which are not of finite type.

Bibliography for Chapter 2

[BL] R. Bowen et O. E. Lanford, "Zeta functions of restrictions of the shift transformation", *Proc. Symp. Pure Math.*, vol **14** AMS, 1970, p. 907-918.

[Bow] R. Bowen, "On Axiom A diffeomorphisms", *Regional Conference Series in Mathematics*, AMS, 1978.

[Bro] J. R. Brown, "Ergodic theory and topological dynamics", Academic Press, 1976.

[CP] E. M. Coven, M. E. Paul, "Sofic systems", *Israel J. of Math.* **20**, 1975, pp. 165-177.

[DGS] M. Denker, C. Grillenberger et K. Sigmund, "Ergodic theory on compact spaces", Lecture Notes in Mathematics, **66 – 67**, S.M.F. (1979).

[FLP] A. Fathi, F. Laudenbach and V. Poénaru, "Travaux de thurston sur les surfaces", Astérisque, vol. **527**, Springer Verlag.

[Fis] R. Fischer, "Sofic systems and graphs", *Monatsh. Math.* **80**, 1975, pp. 179-186.

[Fri] D. Fried, "Finitely presented dynamical systems", *Erg. th. & Dyn. Syst.* **7**, 1987, pp. 489-507.

[Gro 1] M. Gromov, "Hyperbolic manifolds, groups and actions", *Ann. of Math. Studies* **97**, Princeton University Press, 1982 , pp. 183-215.

[Gro 3] ——, "Hyperbolic groups", *in* Essays in Group Theory, MSRI publ. **8**, Springer, 1987, pp. 75-263.

[Had] J. Hadamard, "Les surfaces à courbures opposées et leurs lignes géodésiques", *J. de Math. pures et appl.* **4**, 1898, pp. 27-74.

[Hed] G. A. Hedlund, "Endomorphisms and automorphisms of the shift dynamical system", *Math. Syst. Theory* **3**, 1969, pp. 320-375.

[HW] W. Hurewicz, H. Wallman, "Dimension theory", Princeton University Press, 1948.

[Kur] K. Kuratowski, "Introduction to set theory and topology", 2nd edition, Pergamon Press, 1972.

[Man] A. Manning, "Axiom A diffeomorphisms have rational zeta functions", *Bull.*

Lon. Math. Soc. **3**, 1971, pp. 215-220.

[Moi] E.E. Moise, "Geometry and topology in dimensions 2 and 3", Springer Verlag, 1977.

[Sma] S. Smale, "Differentiable dynamical systems", *Bull. Amer. Math. Soc.* **73**, 1967, pp. 747-817.

[Wei] B. Weiss, "Subshifts of finite type and sofic systems", *Monatsh. Math.* **77**, 1973, pp. 462-474.

[Wal] P. Walters, "Ergodic theory - Introductory lectures", Lecture Notes in Mathematics, vol. **458**, Springer Verlag.

Chapter 3

The boundary of a hyperbolic group
as a finitely presented dynamical system

Let X be a geodesic space which is proper and δ−hyperbolic. We develop in this chapter a construction due to Gromov which describes the boundary ∂X of X in terms of equivalence classes of "differentials", or "$1 - cocycles$", on the space X.

An equivalent formulation realizes the points of ∂X as equivalence classes of functions on X which have certain properties of the function "distance to a point" or "distance to a convex set". We can associate to such a function "gradient lines". All the gradient lines associated to a given function will converge to a well-defined point of ∂X, and two such functions will be considered as equivalent if and only if their associated gradient lines converge to the same point at infinity. This construction is analogous to the one which describes the boundary ∂X in terms of Busemann functions, or horofunctions on X, in the case where X is a simply conected Riemannian manifold of negative curvature (cf. [BGS] or [Gro 1]).

Suppose now that X is the Cayley graph of a hyperbolic group Γ. A description of the boundary $\partial \Gamma = \partial X$ as a quotient of the space Φ_0 of 1-cocycles on X with integer coefficients will give a finite presentation $\pi_0 : \Phi_0 \to \partial \Gamma$ of the dynamical system $(\partial \Gamma, \Gamma)$ (in the sense defined in Chapter 2). Furthermore, the map π_0 is finite-to-one, that is, there exists an integer N such that $card(\pi_0^{-1}(\xi)) \leq N$, for every $\xi \in \partial \Gamma$.

As a corollary, we get that for every geodesic space X which is proper and δ-hyperbolic and which admits a group Γ which acts isometrically, properly discontinuously and cocompactly, the dynamical system $(\partial X, \Gamma)$ is finitely presented.

§1 – The cocycles φ

Let X be a geodesic proper δ-hyperbolic space. We denote by $B_R(x)$ the closed ball of radius R centered at the point $x \in X$.

Let us fix a real number $d \geq 30\delta + 1$. Consider the set Φ of functions

$$\varphi : \{(x, y) \in X \times X \text{ such that } \mid x - y \mid \leq 3d\} \rightarrow \mathbb{R}$$

which satisfy conditions (i) to (iii) below:

(i) *(the cocycle condition)* $\varphi(x, y) = \varphi(x, z) + \varphi(z, y)$ for all x , y and z such that $diam\{x, y, z\} \leq 3d$.

To every function $\varphi \in \Phi$ and to every point $x \in X$, we associate a function $\bar{\varphi}_x$ defined on the ball $B_{3d}(x)$, by the formula:

$$\bar{\varphi}_x(y) = -\varphi(x, y).$$

Condition (ii) can be stated now:

(ii) *(quasi-convexity)* For all points x, y_0 and y_1 of X such that $\mid x - y_0 \mid \leq 2d$ and $\mid x - y_1 \mid \leq 2d$, for every geodesic segment $[y_0, y_1]$ joining the points y_0 and y_1 and for every $t \in [0, 1]$, we have

$$\bar{\varphi}_x(y_t) \leq t\bar{\varphi}_x(y_0) + (1 - t)\bar{\varphi}_x(y_1) + 4\delta,$$

where y_t is the unique point on $[y_0, y_1]$ satisfying $\mid y_0 - y_t \mid = t \mid y_0 - y_1 \mid$. (Note that for every $t \in [0, 1], y_t$ is an element of $B_{2d+4\delta}(x) \subset B_{3d}(x)$.)

(iii) For every given x, we have, for all $t \in [-d, d]$ and for all $y \in B_d(x)$ such that $\bar{\varphi}_x(y) \geq t$,
$$\bar{\varphi}_x(y) = t + dist(y, \bar{\varphi}_x^{-1}(]-\infty, t])).$$

Let us note a few more properties of the functions φ, which follow immediately from the above three properties:

(iv) For all $x \in X$, we have

$$\varphi(x, x) = \bar{\varphi}_x(x) = 0.$$

Indeed, by property (i), we have $\varphi(x, x) = \varphi(x, x) + \varphi(x, x)$.

(v) $\varphi(x, y) = -\varphi(y, x)$ for all x and y such that $\mid x - y \mid \leq 3d$. Indeed, we have $\varphi(x, x) = 0 = \varphi(x, y) + \varphi(y, x)$.

(vi) $\mid \varphi(x,y) \mid \leq \mid x - y \mid$ for all x and y such that $\mid x - y \mid \leq d$. Indeed, according to (v), we can suppose without loss of generality that $\varphi(x,y) = \bar{\varphi}_x(y) \geq 0$. Property (iii), implies therefore

$$\bar{\varphi}_x(y) = 0 + dist\big(y, \bar{\varphi}_x^{-1}(] - \infty, 0])\big),$$

hence $\bar{\varphi}_x(y) \leq \mid y - x \mid$.

$(vi)'$ Using properties (vi) and (i), we have

$$\mid \varphi(x,y) - \varphi(x',y') \mid \leq \mid x - x' \mid + \mid y - y' \mid$$

for every x, y, x' and y' such that $diam\{x,y,x',y'\} \leq d$. In particular, φ is continuous.

(vii) For every $t \in [0, d]$ and for every $x \in X$, there exists $y \in X$ such that

$$\mid x - y \mid = \varphi(x,y) = t.$$

Indeed, applying (iii) with $y = x$, we have $dist(x, \bar{\varphi}_x^{-1}(] - \infty, -t])) = t$. Let y be a projection of x on the set $\bar{\varphi}_x^{-1}(] - \infty, -t])$. (Recall that a *projection* of a point x on a subset $Y \subset X$ is a point $y \in Y$ such that $\mid x - y \mid = dist(x, Y)$. If Y is a non-empty closed subset of X, then every point of X admits a projection on Y (this is a consequence of the fact that the closed balls in X are compact, because X is proper).) We have $\bar{\varphi}_x(y) \leq -t$. As $\mid x - y \mid = t$, we have necessarily $\varphi(x,y) = t$, by (vi).

§2 – Integration of the cocycles φ

Consider the simplicial complex $P_{3d}(X)$ associated to X (*cf.* Chapter 1, §8). A cocycle φ defines a simplicial 1–cochain on that complex. Indeed, note first that φ defines a function on the set of oriented edges of $P_{3d}(X)$ (we recall that, by definition, these edges are ordered pairs of points of X whose distance is $\leq 3d$). Property (i) of the functions φ shows that this cochain is closed. As $P_{3d}(X)$ is contractible, this cochain is a coboundary. Let $\bar{\varphi}$ be a 0–cochain whose coboundary is φ. Then, $\bar{\varphi}$ is a function defined on the vertices of $P_{3d}(X)$. But the set of vertices is the set X itself. Therefore, $\bar{\varphi}$ is a function defined on X. Two 0–cochains whose coboundary is φ differ by a constant. Thus, we have the following

Proposition 2.1. — *Let φ be an element of Φ. Then, there exists a function $\bar{\varphi}$ defined on X such that for every x and $y \in X$ with $\mid x - y \mid \leq 3d$, we have*

$(2.1.1)$ $\quad \varphi(x,y) = \bar{\varphi}(x) - \bar{\varphi}(y).$

Furthermore, $\bar{\varphi}$ is unique up to an additive constant.

If φ is an element of Φ, we say that a function $\bar{\varphi} : X \to \mathbb{R}$ satisfying $(2.1.1)$ for every $x, y \in X$ such that $\mid x - y \mid \leq 3d$ is a *primitive* of φ. Formula $(2.1.1)$ allows us to consider $\varphi \in \Phi$ as being defined on the whole space $X \times X$.

Proposition 2.2. — *Let $\varphi : X \times X \to \mathbf{R}$ be an element of Φ. Then φ satisfies the following properties:*

(I) $\quad \varphi(x,y) = \varphi(x,z) + \varphi(z,y)$ *for every $x,y,z \in X$.*

(V) $\quad \varphi(y,x) = -\varphi(x,y)$ *for every $x,y \in X$.*

(VI) $\quad | \varphi(x,y) | \leq | x - y |$ *for every $x,y \in X$.*

$(VI)'$ $\quad | \varphi(x,y) - \varphi(x',y') | \leq | x - x' | + | y - y' |$ *for every $x,y,x',y' \in X$.*

PROOF. (I) and (V) follow immediately from formula $(2.1.1)$ which defines φ on $X \times X$. To prove (VI), it is sufficient to consider, on a geodesic segment joining x to y, a sequence of points x_i $(i = 0,1,2,...n)$ in that order, such that $x_0 = x$, $x_n = y$ and $| x_i - x_{i+1} | \leq d$ for every $i = 0,1,...,n-1$. We have then, using (I),

$$\varphi(x,y) = \sum_{i=0}^{n-1} \varphi(x_i, x_{i+1}),$$

and therefore, using the triangle inequality and property (v) of §1,

$$| \varphi(x,y) | \leq \sum_{i=0}^{n-1} | \varphi(x_i, x_{i+1}) |$$

$$\leq \sum_{i=0}^{n-1} | x_i - x_{i+1} | = | x - y |.$$

Property (VI) is an immediate consequence of (I) and (VI). $\blacksquare$

Remark. Property (VI) expresses the fact that any primitive $\bar{\varphi}$ of φ is 1-Lipschitz, that is to say, it verifies $| \bar{\varphi}(x) - \bar{\varphi}(y) | \leq | x - y |$ for every $x,y \in X$.

§3 – Cocycles associated to Busemann functions

Let us begin this section by recalling the definition of the Busemann function h_r associated to a geodesic ray r of X, and by establishing a few properties of h_r which will be useful to see that the function $\varphi(x,y) = h_r(x) - h_r(y)$ belongs to the set Φ.

Let r be a geodesic ray in X, that is to say, a map $r : [0,\infty[\to X$ such that $| r(t) - r(t') | = | t - t' |$ for every $t,t' \geq 0$. The *Busemann function* (or *horofunction*) associated to r is the function $h_r : X \to \mathbf{R}$ defined as

$$h_r(x) = \lim_{t \to \infty} (| x - r(t) | - t).$$

Thus, a Busemann function is a limit of normalized distance functions. This limit exists and is finite. Indeed, the triangle inequality shows that $| x - r(t) | - t$ is a non-inecreasing function of t which is bounded below by $- | x - r(0) |$.

Proposition 3.1. — *The function h_r is $1-$Lipschitz, that is to say, we have*

$$\forall x, y \in X, \; | \, h_r(x) - h_r(y) \, | \leq | \, x - y \, | .$$

PROOF. The triangle inequality gives

$$| \, (| \, x - r(t) \, | - t) - (| \, y - r(t) \, | - t) \, | =$$

$$| \, (| \, x - r(t) \, |) - (| \, y - r(t) \, |) \, | \leq | \, x - y \, | .$$

Making t tend to infinity proves the proposition.   ∎

Recall that a function $f : X \to \mathbb{R}$ is said to be $\epsilon - convex$ ($\epsilon \geq 0$) if, for all points y_0, y_1 in X, for every geodesic segment $[y_0, y_1]$ joining these points and for every $\alpha \in [0, 1]$, we have

$$f(y_\alpha) \leq \alpha f(y_0) + (1 - \alpha) f(y_1) + \epsilon,$$

where y_α is the unique point on $[y_0, y_1]$ satisfying $| \, y_0 - y_\alpha \, | = \alpha \, | \, y_0 - y_1 \, |$.

Lemma 3.2. — *For every $x \in X$, the function $f : X \to \mathbb{R}$ defined by $f(y) = | \, x - y \, |$ is $4\delta-convex$.*

PROOF. When X is a tree ($\delta = 0$), the result follows from the convexity of the function $g(\alpha) = f(y_\alpha)$, $\alpha \in [0, 1]$. In fact, it is easy to see in this case that g is a piecewise linear function whose derivative is -1 on $[0, \alpha_0]$ and 1 on $[\alpha_0, 1]$, where y_{α_0} is the projection of x on $[y_0, y_1]$.

Let us pass now to the general case (δ arbitrary). Let $Z = [y_0, y_1] \cup \{x\}$. By the theorem of approximation by trees (*cf.* Chapter 1, §5), there exists a tree T and a map $F : Z \to T$ such that

(3.2.1) $\forall z, z' \in Z, | \, z - z' \, | - 4\delta \leq | \, F(z) - F(z') \, | \leq | \, z - z' \, |,$

and

(3.2.2) $\forall z, z' \in [y_0, y_1], | \, F(z) - F(z') \, | = | \, z - z' \, | .$

Therefore, we have:

$$| \, x - y_\alpha \, | \leq | \, F(x) - F(y_\alpha) \, | + 4\delta, \text{ from (3.2.1)},$$

$$\leq \alpha \, | \, F(x) - F(y_0) \, | + (1 - \alpha) \, | \, F(x) - F(y_1) \, | + 4\delta,$$

from (3.2.2) and the $0-$convexity of the function "distance to $F(x)$" in the tree T, and

$$\leq \alpha \, | \, x - y_0 \, | + (1 - \alpha) \, | \, x - y_1 \, | + 4\delta$$

using (3.2.1).   ∎

Proposition 3.3. — *The function h_r is $4\delta-$convex.*

PROOF. For every $t \geq 0$, we can write, using the preceding lemma:

$$| y_\alpha - r(t) | \leq \alpha \mid y_0 - r(t) \mid +(1 - \alpha) \mid y_1 - r(t) \mid +4\delta.$$

Therefore, we have:

$$| y_\alpha - r(t) \mid -t \leq \alpha(\mid y_0 - r(t) \mid -t) + (1 - \alpha)(\mid y_1 - r(t) \mid -t) + 4\delta.$$

By letting t tend to infinity, we obtain:

$$h_r(y_\alpha) \leq \alpha h_r(y_0) + (1 - \alpha)h_r(y_1) + 4\delta.$$

∎

Proposition 3.4. — *Let λ be a real number, and x a point in X such that $h_r(x) \geq \lambda$. Then there exists a point p in X such that*

$$\mid x - p \mid = h_r(x) - \lambda \quad and \quad h_r(p) = \lambda.$$

PROOF. For every real number t, consider a geodesic segment $[x, r(t)]$ between x and $r(t)$. For t large enough, let $p(t)$ be the unique point on $[x, r(t)]$ such that $\mid x - p(t) \mid = h_r(x) - \lambda$. We have:

$$(3.4.1) \quad \mid x - r(t) \mid -t = h_r(x) - \lambda + (\mid p(t) - r(t) \mid -t).$$

By hypothesis, X is proper. Consequently, the closed ball centered at x and of radius $h_r(x) - \lambda$ is compact. Therefore, we can find a sequence t_i which tends to infinity and such that the sequence $(p(t_i))$ converges to a point p of X. By passing to the limit, we obtain $\mid x - p \mid = h_r(x) - \lambda$ and, using (3.4.1), $h_r(p) = \lambda$. ∎

Corollary 3.5. — *For every real number λ and for every $x \in X$ such that $h_r(x) \geq \lambda$, we have:*

$$h_r(x) = \lambda + dist(x, h_r^{-1}(] - \infty, \lambda])).$$

PROOF. If y is a point in $h_r^{-1}(] - \infty, \lambda])$, we have, by Proposition 3.1,

$$\mid h_r(x) - h_r(y) \mid \leq \mid x - y \mid .$$

Hence

$$h_r(x) \leq \lambda + dist(x, h_r^{-1}(] - \infty, \lambda])).$$

On the other hand, Proposition 3.4 gives

$$dist(x, h_r^{-1}(] - \infty, \lambda])) \leq h_r(x) - \lambda,$$

and this proves the corollary.   ∎

Let us prove now the following

Proposition 3.6. — *The function $\varphi(x,y)$, defined by*

$$\varphi(x, y) = h_r(x) - h_r(y) \text{ for all } x, y \in X \text{ such that } | x - y | \leq 3d$$

belongs to Φ. Furthermore, h_r is a primitive of φ.

PROOF. We must show that φ satisfies properties (i), (ii) and (iii) of §1. The verification of (i) is immediate. (ii) results from Proposition 3.3. Let us prove (iii) (which is not an immediate consequence of Corollary 3.5 !). Let $x \in X$ and let $\bar{\varphi}_x$ be the function on $B_{3d}(x)$ defined by

$$\bar{\varphi}_x(z) = -\varphi(x, z) = h_r(z) - h_r(x).$$

Let $y \in B_d(x)$, and $t \in [-d, d]$ be such that $\bar{\varphi}_x(y) \geq t$, or, in other words,

$$h_r(y) \geq h_r(x) + t.$$

From Proposition 3.4, there exists a point p of X such that

$$| y - p | = h_r(y) - h_r(x) - t$$

and

$$h_r(p) = h_r(x) + t.$$

Using the fact that the function h_r is 1-Lipschitz (Proposition 3.1), we have

$$h_r(y) - h_r(x) \leq | x - y |$$

which implies that

$$| y - p | \leq 2d.$$

Using the triangle inequality, we have therefore

$$| x - p | \leq | x - y | + | y - p | \leq 3d.$$

As $p \in B_{3d}(x), \bar{\varphi}_x(p) = t$ and $| y - p | = \bar{\varphi}_x(y) - t$, we obtain

$$dist(y, \bar{\varphi}_x^{-1}(] - \infty, t])) \leq \bar{\varphi}_x(y) - t.$$

On the other hand, if $z \in \bar{\varphi}_x^{-1}(] - \infty, t])$, we have

$$\bar{\varphi}_x(y) - t \leq \bar{\varphi}_x(y) - \bar{\varphi}_x(z) = | h_r(y) - h_r(z) | \leq | y - z |.$$

This implies

$$\bar{\varphi}_x(y) - t \leq dist(y, \bar{\varphi}_x^{-1}(] - \infty, t]).$$

49

Thus, we have established that

$$dist(y, \bar{\varphi}_x^{-1}(]-\infty, t])) = \bar{\varphi}_x(y) - t,$$

that is to say, property (iii). ∎

§4 – The gradient lines defined by φ

Let φ be an element of Φ.

Definition 4.1. — A *gradient line for φ*, or a *φ-gradient line* is a rectifiable map g from an interval $I \subset \mathbf{R}$ to X parametrized by arclength and satisfying

$$\forall\, t, t' \in I, \varphi(g(t), g(t')) = t' - t.$$

Remark that a local gradient line is also a global gradient line, as can be seen from the cocycle property (I) of φ given in Proposition 2.2. This cocycle property shows also that if we concatenate two gradient lines (which have the property that the endpoint of the first one equal to the initial point of the second), then the resulting path (parametrized by arclength) is also a gradient line.

Let us begin by proving the following

Proposition 4.2. — *If $g : I \to X$ is a gradient line for $\varphi \in \Phi$, then g is geodesic, i.e.*

$$\forall\, t, t' \in I, |\, g(t) - g(t') \,| = |\, t - t' \,| \,.$$

PROOF. Given t and t' in I with $t \geq t'$, let $g_{[t,t']}$ be the subpath in g comprised between $g(t)$ and $g(t')$. Let us show that this path is geodesic.

As the path g is parametrized by arclength, we have $length(g_{[t,t']}) = t' - t$. On the other hand, we have

$$t' - t = \varphi(g(t), g(t')) \leq |\, g(t') - g(t) \,|$$

(by property (VI) of the functions φ). Therefore, we have

$$length(g_{[t,t']}) = |\, g(t') - g(t) \,|,$$

and therefore the path g is geodesic. ∎

Proposition 4.3. — *Let $\varphi \in \Phi$ and $x \in X$, let $y \in X$ be a point satisfying $\varphi(x, y) = |\, x - y \,|$, and let $g : [t_0, t_1] \to X$ be a geodesic segment between x and y. Then, g is a gradient line for φ.*

PROOF. By property (VI) of the functions φ, we have, for every $t' \geq t$,

$$\varphi(g(t), g(t')) \leq | g(t) - g(t') | = t' - t.$$

Suppose that there exists $t' \geq t$ such that $\varphi(g(t), g(t')) < t' - t$. We would have then (again by application of property (VI))

$$t_1 - t_0 = \varphi(x, y) = \varphi(g(t_0), g(t)) + \varphi(g(t), g(t')) + \varphi(g(t'), g(t_1))$$

$$< (t - t_0) + (t' - t) + (t_1 - t') = t_1 - t_0,$$

which is a contradiction. Therefore, we have $\varphi(g(t), g(t')) = t' - t$ for every $t, t' \in [t_0, t_1]$. ∎

Proposition 4.4. — *Let φ be an element of Φ and x a point in X. Then, there exists a φ-gradient line $g : [0, \infty[\to X$ starting at x, that is, such that $g(0) = x$.*

PROOF. From property (vii), we can construct, by induction, a sequence of points x_i, $i = 0, 1, ...$, such that $x_0 = x$ and such that, for every i,

$$\varphi(x_i, x_{i+1}) = | x_i - x_{i+1} | = d.$$

Consider, for every i, a geodesic segment $[x_i, x_{i+1}]$ between x_i and x_{i+1}. Let $g : [0, \infty[\to X$ be the infinite path obtained by concatenating the paths $[x_i, x_{i+1}]$, parametrized by arclength, starting at $x = x_0 = g(0)$. By Proposition 4.3, each of the paths $[x_i, x_{i+1}]$ is a gradient line. By the cocycle condition (I), the path g is a φ-gradient line (*cf.* the remark following Definition 4.1). ∎

The preceding proposition provides, for the cocycles $\varphi \in \Phi$, "global" versions of properties (iii) and (vii) of §1:

Corollary 4.5. — *Let φ be an element of Φ. Then:*

(VII) *For every $x \in X$ and for every real number $t \geq 0$, there exists $y \in X$ such that*

$$\varphi(x, y) = | x - y | = t.$$

PROOF. Let $g : [0, \infty[\to X$ be a φ-gradient line starting at x. The point $y = g(t)$ does the job. ∎

Corollary 4.6. — *Let $\varphi \in \Phi$ and $\bar{\varphi}$ a primitive of φ. Then, for every $x \in X$ and for every real number t such that $\bar{\varphi}(x) \geq t$, we have*

(III) $\bar{\varphi}(x) = t + dist(x, \bar{\varphi}^{-1}(] - \infty, t]))$.

PROOF. This is an immediate consequence of the preceding corollary and of property (VI) of φ. ∎

§5 – The point at infinity associated to a cocycle

Given an element $\varphi \in \Phi$, let us choose a point $x \in X$ and a gradient line g starting at x. As this line is geodesic, it converges to a well-defined point of ∂X. We will show that this point is canonically associated to φ (that is, it does not depend on x or on the chosen gradient line). We shall denote this point by $a(\varphi)$, and we call it *the point at infinity associated to φ.*

Proposition 5.1. — *Let $\varphi \in \Phi$, and let $g, g' : [0, \infty[\to X$ be two φ-gradient lines starting respectively at the points x and y of X. Then, g and g' stay at a uniformly bounded distance from each other. In other words, there exists a constant C such that*

$$\forall t \geq 0, |\, g(t) - g'(t)\, | \leq C.$$

PROOF. The proof will be done in three steps:

First step. Let us show that the statement is true with the assumption that $\varphi(x, y) = 0$ and $|\, x - y\, | \leq 26\delta + 1$. Let (x_i) (resp. (y_i)) be the sequence of points on g (resp. on g') defined by $x_i = g(di)$ (resp. $y_i = g'(di)$), $i = 0, 1, 2, \dots$ It suffices to show that the two sequences x_i and y_i stay at a uniformly bounded distance from each other. In fact, we shall prove by induction that for every i, we have

$$|\, x_i - y_i\, | \leq 26\delta + 1.$$

This is true, by hypothesis, for $i = 0$. Suppose that $|\, x_i - y_i\, | \leq 26\delta + 1$ for a given i. Let $[x_{i+1}, y_{i+1}]$ be a geodesic segment joining the points x_{i+1} and y_{i+1}. Consider the set $Z = [x_{i+1}, y_{i+1}] \cup \{x_i, y_i\}$. From the theorem of approximation by trees, there exists a tree T and a map $f : Z \to T$ with the following properties:

(5.1.1) $\quad \forall z_1, z_2 \in Z, |\, z_1 - z_2\, | - 6\delta \leq |\, f(z_1) - f(z_2)\, | \leq |\, z_1 - z_2\, |,$
and

(5.1.2) $\quad \forall z_1, z_2 \in [x_{i+1}, y_{i+1}], |\, f(z_1) - f(z_2)\, | = |\, z_1 - z_2\, |.$

On $B_{3d}(x_i) \cap B_{3d}(y_i)$, we have $\bar{\varphi}_{x_i} = \bar{\varphi}_{y_i}$, because $\varphi(x_i, y_i) = 0$ by the cocycle condition. The point x_{i+1} (resp. y_{i+1}) is a projection of x_i (resp. y_i) on the set $\bar{\varphi}_{x_i}^{-1}(]-\infty, -d])$.

By $\bar{\varphi}_{x_i}(x_{i+1}) = \bar{\varphi}_{x_i}(y_{i+1}) = -d$, we have, using property *(ii)* of the functions φ (quasi-convexity):

$$\forall z \in [x_{i+1}, y_{i+1}], \bar{\varphi}_{x_i}(z) \leq -d + 4\delta.$$

(Remark that we have $x_{i+1}, y_{i+1} \in B_{2d}(x_i)$ because $|\, x_i - x_{i+1}\, | = d$ and $|\, x_i - y_{i+1}\, | \leq |\, x_i - y_i\, | + |\, y_i - y_{i+1}\, | \leq 26\delta + 1 + d \leq 2d$.)

As $-d + 4\delta \leq 0$, property *(iii)* of the functions φ implies, for every $z \in [x_{i+1}, y_{i+1}]$,

$$|\, z - x_i\, | \geq d - 4\delta \quad \text{and} \quad |\, z - y_i\, | \geq d - 4\delta.$$

Therefore, using (5.1.1), we have

(5.1.3) $\mid p - f(x_i) \mid \geq d - 10\delta$ and $\mid p - f(y_i) \mid \geq d - 10\delta$,

for every point p on the segment $[f(x_{i+1}), f(y_{i+1})]$ which is the image of $[x_{i+1}, y_{i+1}]$ by f, as follows from (5.1.2).

Consider now, in the tree T, the projections p_i and q_i respectively of the points $f(x_i)$ and $f(y_i)$ on the segment $[f(x_{i+1}), f(y_{i+1})]$. If p_i and q_i are distinct (Figure 1), we have

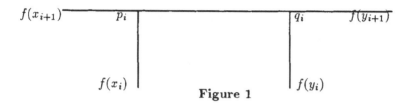

Figure 1

$$\mid x_i - y_i \mid \geq \mid f(x_i) - f(y_i) \mid \text{ from } (5.1.1),$$
$$\geq \mid f(x_i) - p_i \mid + \mid f(y_i) - q_i \mid$$
$$\geq 2d - 20\delta, \text{ using } (5.1.3),$$

which contradicts the induction hypotesis $\mid x_i - y_i \mid \leq 26\delta + 1$. Therefore, $f(x_i)$ and $f(y_i)$ have the same projection $p = f(z)$ on the segment $[f(x_{i+1}), f(y_{i+1})]$ as shown in Figure 2.

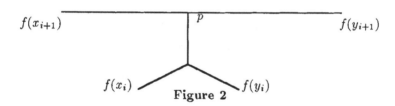

Figure 2

Again by (5.1.3), we have

$$\mid f(x_i) - p_i \mid \geq d - 10\delta.$$

On the other hand, we know that

$$\mid f(x_i) - f(x_{i+1}) \mid \leq \mid x_i - x_{i+1} \mid = d.$$

This implies that

$$\mid p - f(x_{i+1}) \mid \leq 10\delta.$$

In the same manner, we can show that

$$| p - f(y_{i+1}) | \leq 10\delta.$$

Finally, we have

$$| f(x_{i+1}) - f(y_{i+1}) | = | p - f(x_{i+1}) | + | p - f(y_{i+1}) | \leq 20\delta,$$

and therefore, using (5.1.1),

$$| x_{i+1} - y_{i+1} | \leq 26\delta.$$

Second step. Let us show that the statement is true under the following assumption:

$$| x - y | \leq 13\delta + 1/2.$$

At the expense of possibly interchanging x and y, we can suppose that $\varphi(x, y) \geq 0$. Let $x_0 = g(t_0)$ where $t_0 = \varphi(x, y)$. We have

$$\varphi(x_0, y) = \varphi(x_0, x) + \varphi(x, y) = 0.$$

On the other hand,

$$| x - x_0 | = t_0 = \varphi(x, y) \leq | x - y | \leq 13\delta + 1/2.$$

Using the triangle inequality, we obtain

$$| x_0 - y | \leq | x - x_0 | + | x - y | \leq 26\delta + 1.$$

Applying the result of the first step to the gradient lines $(g(t_0 + t))_{t \geq 0}$ and $(g'(t))_{t \geq 0}$, we see that these two lines stay at a uniformly bounded distance from each other. The triangle inequality shows that the same thing holds for g and g'.

Third step. Let us show now that the statement is true under the most general assumptions. The space X being geodesic, we can find a finite sequence of points x_i ($i = 0, 1, ..., n$) such that $x_0 = x, x_n = y$ and $| x_i - x_{i+1} | \leq 13\delta + 1/2$ for every $i = 0, 1, ..., n - 1$. Consider a sequence of gradient lines $g_i : [0, \infty[\to X$ ($i = 0, 1, ..., n$) such that $g_i(0) = x_i$ for every i, $g_0 = g$ and $g_n = g'$. Applying the result of the second step to (g_i, g_{i+1}) for $i = 0, 1, ..., n - 1$ and using the triangle inequality, we see that the gradient lines g and g' stay at a uniformly bounded distance from each other. ∎

Corollary 5.2. — *Given an element $\varphi \in \Phi$, a point $x \in X$ and a φ-gradient line $g : [0, \infty[\to X$ starting at x, the associated point at infinity $g(\infty) \in \partial X$ depends only on φ. We denote this point by $a(\varphi)$.* ∎

Definition 5.3. — We say that two elements φ_1 and φ_2 of Φ are *equivalent* ($\varphi_1 \sim \varphi_2$) if the following is satisfied:

$$a(\varphi_1) = a(\varphi_2).$$

The next proposition will be useful:

Proposition 5.4. — *Consider a geodesic ray r which converges to a point $\xi \in \partial X$, an infinite sequence $x_0, x_1, x_2, \dots$ of points on r converging to ξ and, for every $i = 0, 1, 2, \dots$, the closed ball B_i centered at x_i and of radius 8δ. Let φ_1 and φ_2 be two elements of Φ such that $a(\varphi_1) = a(\varphi_2) = \xi$. Suppose that φ_1 and φ_2 have the same restriction on the union of the $B_i \times B_i$. Then, $\varphi_1(x, y) = \varphi_2(x, y)$ for every $(x, y) \in X \times X$.*

PROOF. Let x and y be points in X. Let $g_1 : [0, \infty[\to X$ be a φ_1-gradient line starting at x and $g_2 : [0, \infty[\to X$ a φ_2-gradient line starting at y. We know that g_1 and g_2 are geodesics which converge to ξ. Consider a geodesic segment $[r(0), x]$ and the geodesic triangle $\Delta = [r(0), \xi, x]$ which has r, g_1 and $[r(0), x]$ as sides. By the hyperbolicity of X, we know that Δ is 8δ-narrow: every point on one of its sides is at distance $\leq 8\delta$ from the union of the two other sides (Proposition 3.2 of Chapter 1, with $n = 3$ and $p = 1$). For each i large enough, we have $dist(x_i, [x, r(0)]) > 8\delta$ and the ball B_i has nonempty intersection with g_1. In the same way, B_i has nonempty intersection with g_2 for every i large enough. Let us choose an integer i such that B_i has nonempty intersection with both g_1 and g_2. Let p_1 be a point in $B_i \cap g_1$ and p_2 a point in $B_i \cap g_2$. We have

$$\varphi_2(x, y) = \varphi_2(x, p_1) + \varphi_2(p_1, p_2) + \varphi_2(p_2, y),$$

by the cocycle property (I) satisfied by φ_2. Using property (VI), we have

$$\varphi_2(x, p_1) \leq | x - p_1 | = \varphi_1(x, p_1)$$

and

$$\varphi_2(p_2, y) = - | p_2 - y | \leq \varphi_1(p_2, y).$$

On the other hand, we have $\varphi_2(p_1, p_2) = \varphi_1(p_1, p_2)$, as φ_1 and φ_2 coincide on $B_i \times B_i$. It follows that

$$\varphi_2(x, y) \leq \varphi_1(x, p_1) + \varphi_1(p_1, p_2) + \varphi_1(p_2, y) = \varphi_1(x, y).$$

In the same way, $\varphi_1(x, y) \leq \varphi_2(x, y)$. Hence $\varphi_1(x, y) = \varphi_2(x, y)$, which proves the proposition. ∎

§6 – Properties of the map $a : \Phi \to \partial X$

We introduce now the metric $| \ |_d$ on Φ, defined by

$$| \varphi_1 - \varphi_2 |_d = \sup_{|x-y| \leq d} | \varphi_1(x, y) - \varphi_2(x, y) |.$$

We have $| \varphi_1 - \varphi_2 |_d < \infty$, using the fact that $| \varphi(x, y) | \leq d$ for all $\varphi \in \Phi$ and for all $x, y \in X$ such that $| x - y | \leq d$. Let us note also that if $| \varphi_1 - \varphi_2 |_d = 0$, then

$\varphi_1(x, y) = \varphi_2(x, y)$ for every $(x, y) \in X \times X$ (to see this, it suffices to join x and y by a chain of points of X whose mesh is $\leq d$ and use the cocycle condition). In fact, the metric $\mid \;\mid_d$ on Φ is the restriction of the metric associated to the *sup* norm on the vector space of bounded functions on $\{(x, y) \in X \times X; \mid x - y \mid \leq d\}$. Here, as in the definition of the set Φ, we prefer to give "local" definitions (i. e. for x and y close enough) rather than global ones, which we could have also taken, the local definitions being useful for the theory of finitely presented dynamical systems in which we are interested.

Proposition 6.1. — *Let φ_1 and φ_2 be two elements of Φ such that $a(\varphi_1) = a(\varphi_2)$. Then, for every $x, y \in X$, $\mid \varphi_1(x, y) - \varphi_2(x, y) \mid \leq 24\delta$.*

PROOF. Let x and y be two points of X, let g_1 and g_1' be two φ_1–gradient lines starting respectively at x and y, and let g_2 and g_2' be two φ_2–gradient lines starting at these same points (Figure 3).

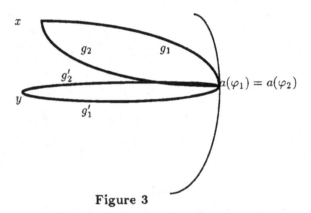

Figure 3

By a property of geodesic rays starting at the same point and converging to the same point at infinity (see inequality 3.4.1 of Chapter 1), we have $\mid g_1(t) - g_2(t) \mid \leq 4\delta$ and $\mid g_1'(t) - g_2'(t) \mid \leq 4\delta$ for every $t \geq 0$.

On the other hand, for every t which is large enough, the point $g_2(t)$ is contained in the 8δ–neighborhood of the geodesic g_1' (Corollary 3.5 of Chapter 1). Thus, we can find two real numbers $t_0 \geq 0$ and $t_0' \geq 0$ such that $\mid g_1(t_0) - g_2(t_0) \mid \leq 4\delta$, $\mid g_2(t_0) - g_1'(t_0') \mid \leq 8\delta$ and $\mid g_1'(t_0') - g_2'(t_0') \mid \leq 4\delta$.

Let now $\bar{\varphi}_1$ and $\bar{\varphi}_2$ be primitives of φ_1 and φ_2 respectively. We have:

$$\varphi_1(x, y) - \varphi_2(x, y)$$

$$= (\bar{\varphi}_1(x) - \bar{\varphi}_1(y)) - (\bar{\varphi}_2(x) - \bar{\varphi}_2(y))$$

$$= (\bar{\varphi}_1(x) - t_0) - (\bar{\varphi}_1(y) - t_0') - ((\bar{\varphi}_2(x) - t_0) - (\bar{\varphi}_2(y) - t_0'))$$

$$= (\bar{\varphi}_1(g_1(t_0)) - \bar{\varphi}_1(g_1'(t_0'))) - (\bar{\varphi}_2(g_2(t_0)) - \bar{\varphi}_2(g_2'(t_0'))).$$

Using the fact that the functions $\bar{\varphi}_1$ and $\bar{\varphi}_2$ are 1–Lipschitz, we obtain finally:

$$| \varphi_1(x,y) - \varphi_2(x,y) | \leq 12\delta + 12\delta = 24\delta.$$

∎

Proposition 6.2. — *Let φ_1 and φ_2 be two elements of Φ with $a(\varphi_1) \neq a(\varphi_2)$, and let γ be a geodesic in X with endpoints $a(\varphi_1)$ and $a(\varphi_2)$. Then, for all points x and y on γ, we have:*

$$| \varphi_1(x,y) - \varphi_2(x,y) | \geq 2 | x - y | -8\delta.$$

PROOF. Suppose that γ is parametrized by arclength, with $\gamma(-\infty) = a(\varphi_2)$ and $\gamma(\infty) = a(\varphi_1)$, and let $x = \gamma(0)$ be the initial point of this geodesic.

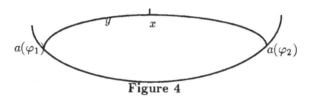

Figure 4

Without loss of generality, we can assume that y is situated on γ between the points x and $a(\varphi_1)$. We have therefore $y = \gamma(t)$ with $t =| x - y |$.

Let $g_1 : [0, \infty[\to X$ be a φ_1-gradient line starting at x. As the geodesic rays $\gamma_{|[0,\infty[}$ and g_1 have the same endpoints, we have $| \gamma(t) - g_1(t) | \leq 4\delta$. Using $(VI)'$, we obtain

$$(6.2.1) \quad | \varphi_1(x,y) - t | \leq 4\delta.$$

In the same way, we can use a φ_2-gardient line g_2 starting at y to obtain

$$(6.2.2) \quad | \varphi_2(y,x) - t | \leq 4\delta.$$

Inequalities (6.2.1) and (6.2.2) give

$$| \varphi_1(x,y) - \varphi_2(x,y) - 2t | \leq 8\delta$$

and therefore, in particular,

$$| \varphi_1(x,y) - \varphi_2(x,y) | \geq 2t - 8\delta$$

which is the desired inequality. ∎

Corollary 6.3. — *For every $\varphi_1, \varphi_2 \in \Phi$ with $a(\varphi_1) \neq a(\varphi_2)$, and for every $t \geq 0$, we can find points $x, y \in X$ with $| x - y |= t$ and*

$$| \varphi_1(x,y) - \varphi_2(x,y) | \geq 2t - 8\delta.$$

∎

Corollary 6.4. — *If* $| \varphi_1 - \varphi_2 |_d \leq 52\delta$, *then* $a(\varphi_1) = a(\varphi_2)$.

PROOF. We have seen that if $a(\varphi_1) \neq a(\varphi_2)$, we can find points x and y with $| x - y | \leq d$, and such that

$$| \varphi_1(x,y) - \varphi_2(x,y) | \geq 2d - 8\delta > 52\delta.$$

∎

Proposition 6.5. — *Let* φ_1 *and* φ_2 *be two elements of* Φ. *We have* $\varphi_1 \sim \varphi_2$ *if and only if* $| \varphi_1 - \varphi_2 |_d \leq 52\delta$.

PROOF. This is an immediate consequence of Proposition 6.1 and of Corollary 6.4. ∎

We equip the set Φ with the topology of uniform convergence on compact subsets of $X \times X$. Let us note that for every $\alpha > 0$, this topology is also the topology of uniform convergence on the compact sets of $\{(x,y) \in X \times X$ such that $| x - y | \leq \alpha\}$ (to see this, use the cocycle condition (I)). The topology of Φ can also be described in the following manner. Equip the set $\bar{\Phi}$ of primitives of elements of Φ with the topology of uniform convergence on compact subsets of X. We can see easily that the topology of Φ is the quotient topology for the canonical map $\bar{\Phi} \to \Phi$.

Remark. One has to be aware of the fact that the topology that we have just defined on Φ (and which is the topology that we shall be using) is *coarser* (i.e. has less open sets) than the topology which is induced by the metric $| \ |_d$.

Let us begin by the following

Proposition 6.6. — *The space* Φ *is compact.*

PROOF. Let $\mathcal{D}$ be the space of continuous functions on

$$\{(x,y) \in X \times X \text{ satisfying } | x - y | \leq 3d\},$$

equipped with the topology of uniform convergence on compact subsets. As a subspace of $\mathcal{D}$, the space Φ is closed. Indeed, properties (i), (ii) and (iii) of the cocycles φ remain true when we take limits. Thus, Φ is a closed subset of equicontinuous functions of $\mathcal{D}$ (as follows from $(vi)'$) whose set of values is bounded. By Ascoli's theorem ($cf.$ [Kel]), Φ is compact. ∎

The group $Isom(X)$ has a continuous left action on Φ, defined by the formula

$$\gamma\varphi(x,y) = \varphi(\gamma^{-1}x, \gamma^{-1}y),$$

for all $\gamma \in Isom(X)$ and $\varphi \in \Phi$. (One can easily verify that $\gamma\varphi$ satisfies, as φ, properties (i), (ii) and (iii) of elements of Φ.)

Proposition 6.7. — *The map $a : \Phi \to \partial X$ is continuous, surjective and $Isom(X)$-equivariant.*

PROOF. For the surjectivity, let ξ be an element of ∂X, and let us take a geodesic ray r which converges to this point. If h_r is the associated Busemann function, we have seen in §3 that the function $\varphi_r(x, y) = h_r(x) - h_r(y)$ defines an element of Φ. The point at infinity which is associated to this cocycle is ξ. Indeed, for every $t \geq 0$, we have $h_r(r(t)) = -t$, and therefore, for every t_1, $t_2 \geq 0$, we have

$$\varphi_r(r(t_1), r(t_2)) = h_r(r(t_1)) - h_r(r(t_2)) = t_2 - t_1,$$

which shows that the ray r is a φ_r−gradient line. Therefore, $a(\varphi_r) = r(\infty) = \xi$.

For the continuity, let φ_n be a sequence of elements of Φ which converges to $\varphi \in \Phi$, and consider a sequence of primitives $\bar{\varphi}_n$ of φ_n, converging to a primitive $\bar{\varphi}$ of φ, the sequences $\bar{\varphi}_n$ and $\bar{\varphi}$ being chosen so that they take the same value at a certain fixed point x of X.

Starting at the point x, we consider gradient lines g_n associated to φ_n for every n, constructed as in the proof of Proposition 4.4, by taking a sequence of successive projections on sets of the form $(\bar{\varphi}_n^{-1}([-\infty, -d]))$. As the values of the functions $\bar{\varphi}_n$ are close from those of $\bar{\varphi}$ uniformly on compact sets of X, we can choose the sequence of projections that are used in the construction of the lines g_n in such a way that they are close to the sequence of projections associated to φ. Thus, the lines g_n, which are geodesic, will be 4δ-close, uniformly on compact sets, from the gradient line associated to φ, and the sequence $a(\varphi_n)$ converges to $a(\varphi)$.

If $g : [0, \infty[\to X$ is a gradient line for $\varphi \in \Phi$ and if $\gamma \in Isom(X)$, we remark that $\gamma \circ g$ is a gradient line for $\gamma\varphi$. We have therefore:

$$a(\gamma\varphi) = \gamma \circ g(\infty) = (\gamma g)(\infty) = \gamma a(\varphi).$$

This shows that the map $a : \Phi \to \partial X$ is $Isom(X)$-equivariant and finishes the proof of Proposition 6.7. ∎

Let us consider now the space $\Phi/ \sim$. This space is equipped with the quotient topology induced from the space Φ. By passing to the quotient, the map a defines a bijection

$$\eta : \Phi/ \sim \to \partial X.$$

The space $\Phi/ \sim$, as Φ itself, is compact (this is a consequence of the fact that the relation $\sim$ is closed, using the the continuity of a). Thus, we have the following

Corollary 6.8. — *The map η is a $Isom(X)$-equivariant homeomorphism between $\Phi/ \sim$ and ∂X.* ∎

§7 – Integral cocycles on a hyperbolic graph

In the rest of this chapter, we suppose that the space X is a graph (*i.e.* a simplicial complex of dimension one) which is connected and locally finite. X is equipped with its canonical metric (see Chapter 1). In this way, X becomes a geodesic space which is proper, and we suppose that, equipped with this metric, X is δ-hyperbolic. Furthermore, we suppose that this graph X has *bounded geometry*, that is to say, there exists an integer M such that the number of edges adjacent to an arbitrary vertex of X is bounded by M.

Let X^0 be the set of vertices of X. This set is equipped with a metric induced by that of X. We shall say that the two vertices x and $y \in X^0$ are *neighboring vertices* if we have $\mid x - y \mid = 1$.

For such a space X, we are mostly interested in the subset Φ_0 of Φ consisting of the cocycles which take integer values on all pairs of points in X^0. It is clear that Φ_0 is a closed subset of Φ and that Φ_0 is invariant under the action of the group $Aut(X) \subset Isom(X)$ of simplicial automorphisms of X. A function φ which belongs to Φ_0 is said to be an *integral cocycle*.

In what follows, we shall make a systematic use of the following

Proposition 7.1. — *Let $x, y \in X^0$ be neighboring vertices, and let $[x, y]$ be the edge joining them. Let $\varphi \in \Phi_0$. Then $\varphi(x, y) = -1, 0$ or 1, and the restriction of φ to $[x, y] \times [x, y]$ is completely determined by $\varphi(x, y)$. (In fact, if $\varphi(x, y) = -1$ (resp. 0, resp. 1), we have $\varphi(x, z) = - \mid x - z \mid$ (resp. $-dist(z, \{x, y\})$, resp. $\mid x - z \mid$), for every $z \in [x, y]$.)*

PROOF. We have $\mid \varphi(x, y) \mid \leq \mid x - y \mid = 1$. Hence, $\varphi(x, y) = -1, 0$ or 1. If $\varphi(x, y) = -1$ or 1, Proposition 4.3 shows that the geodesic segment $[x, y]$, with the appropriate orientation, is a φ-gradient line, which concludes the argument. Suppose now that $\varphi(x, y) = 0$. Consider a φ-gradient line $g : [0, \infty[\to X$ starting at the middle point m of $[x, y]$. This ray g passes necessarily through x or y. Hence, $\varphi(x, m) = -1/2$ and we can conclude the proof by using again Proposition 4.3. ∎

Corollary 7.2. — *An element of Φ_0 is completely determined by its restriction to $X^0 \times X^0$.* ∎

Corollary 7.3. — *The space Φ_0 is totally disconnected.*

PROOF. Given two points x and y in X^0, the map which associates the value $\varphi(x, y)$ to $\varphi \in \Phi_0$ is continuous. As $\varphi(x, y) \in \mathbb{Z}$, it follows that all the elements of a connected subset C of Φ_0 have the same restriction to $X^0 \times X^0$. The preceding corollary implies then $card(C) \leq 1$. ∎

The map $a : \Phi \to \partial X$ defines by restriction a map $a_0 : \Phi_0 \to \partial X$ which has the following property:

Proposition 7.4. — *The map $a_0 : \Phi_0 \to \partial X$ is continuous, surjective and $Aut(X)$-equivariant.*

PROOF. It suffices to prove the surjectivity (the other properties follow from the analogous properties of the map $a : \Phi \to \partial X$).

Let r be a geodesic ray in X, with $r(0) \in X^0$. For every $n = 0, 1, 2, ...$, we have also $r(n) \in X^0$.

For every $x \in X$, we can write

$$h_r(x) = lim_{n \to \infty}(\mid x - r(n) \mid - n).$$

If x is an element of X^0, all the terms of the sequence $\mid x - r(n) \mid - n$ are integers, and $h_r(x)$ is therefore an integer.

Thus, for every $\xi \in \partial X$, we can find a Busemann function whose cocycle is in Φ_0, and whose point at infinity is ξ. This proves the surjectivity of the map a_0, and completes the proof of the proposition. ∎

By restriction of the relation $\sim$ defined in the previous paragraph, there is an equivalence relation on the set Φ_0, which we shall also denote by $\sim$. The map a_0 descends to the quotient and defines a map $\eta_0 : \Phi_0 / \sim \to \partial X$. Proposition 7.4 implies the following

Corollary 7.5. — *The map $\eta_0 : \Phi_0 / \sim \to \partial X$ is an $Aut(X)$-equivariant homeomorphism.* ∎

We shall prove next that the map a_0 is finite-to-one, and for that purpose we need a few lemmas.

Let M be an upper bound for the number of edges which contain an arbitrary vertex of X.

Lemma 7.6. — *Let B be a closed ball in X^0 whose radius is an integer $N_0 \geq 0$. The number of points in B is bounded above by $N_1 = 1 + M + M^2 + ... + M^{N_0}$.*

PROOF. By induction on N_0. ∎

Lemma 7.7. — *Let B be as in the preceding lemma. The number of distinct restrictions of elements $\varphi \in \Phi_0$ to the set $B \times B$ is bounded above by $(2N_0 + 1)^{N_1 - 1}$.*

PROOF. Let x be the center of the ball B. We are looking for an upper bound for the number of functions on B which are of the form $\varphi(x, .)$, where φ is the restriction of an element of Φ_0.

For every $y \in B$, we have $\varphi(x, x) = 0$, $\mid y - x \mid \leq N_0$, and we know that $\varphi(x, .)$ is 1–Lipschitz. The function $\varphi(x, .)$ can take at most $2N_0 + 1$ distinct values on B. As $card\{y \in B \mid y \neq x\} \leq N_1 - 1$, the number of functions $\varphi(x, .)$ under consideration is bounded above by $(2N_0 + 1)^{N_1 - 1}$. ∎

Consider now a point $\xi \in \partial X$ and a geodesic ray r with $r(0) \in X^0$ and $r(\infty) = \xi$. Let $N_0 = [8\delta + 1]$ (integral value), and consider a sequence of consecutive points $x_i \in X^0$, $i = 0, 1, 2, ...$, on the ray r, with $\mid x_i - x_{i+1} \mid > N_0 + 8\delta$ for every $i = 0, 1, 2, ...$

Finally, for every i, let B_i be the closed ball in X^0 centered at x_i and of radius N_0.

Lemma 7.8. — *If φ and φ' are two elements of Φ_0 with $a(\varphi) = a(\varphi') = \xi$, and if for a given $i \in \{1, 2, 3, ...\}$, φ and φ' have the same restriction on $B_i \times B_i$, then these two functions have also the same restriction on $B_{i-1} \times B_{i-1}$.*

PROOF. Let x be a point in B_{i-1} and let g be a φ–gradient line starting at x. Consider the infinite geodesic triangle consisting of the union of g with the subray of r starting at the point x_{i-1} and a geodesic segment $[x_{i-1}, x]$. By Proposition 3.2 of Chapter 1, every point on one of the sides of this triangle is at distance $\leq 8\delta$ from the union of the two other sides. We have $dist(x_i, [x_{i-1}, x]) > 8\delta$. Therefore, there exists a point y on g with $|x_i - y| \leq 8\delta$. The φ–gradient line g has therefore nonempty intersection with the ball B_i.

In the same manner, every φ'–gradient line starting at a point in B_{i-1} has nonempty intersection with the ball B_i.

As the two cocycles φ and φ' take the same values on $B_i \times B_i$, we can prove now, using the same argument than the one in the proof of Poposition 5.4, that the cocycles φ and φ' take the same values on $B_{i-1} \times B_{i-1}$. ∎

We can now prove the following

Proposition 7.9. — *The map $a_0 : \Phi_0 \to \partial X$ is finite-to-one. More precisely, for every point $\xi \in \partial X$, we have $card(a_0^{-1}(\xi)) \leq N = (2N_0 + 1)^{N_1 - 1}$, where, as before, $N_0 = [8\delta + 1]$ and where N_1 is an upper bound for the points of X^0 which are contained in a ball of radius N_0.*

PROOF. Let $\mathcal{F}$ be a finite subset of $a_0^{-1}(\xi)$, and let $N' = card(\mathcal{F})$. Consider a geodesic ray r starting at a vertex of X and converging to ξ, and a sequence of successive balls $B_0, B_1, B_2, ...$ which have the same properties as the balls defined above.

By Proposition 5.4, we know that for two distinct functions φ and φ' in $\mathcal{F}$, there exists a ball B_i such that φ and φ' have different restrictions on $B_i \times B_i$.

By Lemma 7.8, there exists an integer $i_0 \in \{0, 1, 2, ...\}$ such that for every $i \geq i_0$, the restrictions of φ and φ' on $B_i \times B_i$ are distinct.

By making the same consideration for every pair of distinct cocycles φ and φ' in $\mathcal{F}$, we find, taking i large enough, a ball B_i such that the restrictions to $B_i \times B_i$ of the N' cocycles in $\mathcal{F}$ are all distinct.

By Lemma 7.7, we have therefore $N' \leq N = (2N_0 + 1)^{N_1 - 1}$, and the proof of Proposition 7.9 is complete. ∎

§8 – A finite presentation of the boundary of a hyperbolic group

Let Γ be a hyperbolic group and $\partial \Gamma$ its boundary. In this paragraph, we use the preceding results to prove the following

Theorem 8.1. — *There exists a finite set S, a subshift of finite type $\Phi_0 \subset \Sigma(\Gamma, S)$ and a map $\pi_0 : \Phi_0 \to \partial \Gamma$ satisfying the following three properties:*

(*i*) π_0 *is continuous, surjective and Γ-equivariant.*

(*ii*)   *The equivalence relation*

$$R(\pi_0) = \{(\varphi_1, \varphi_2) \mid \pi_0(\varphi_1) = \pi_0(\varphi_2)\} \subset \Sigma(\Gamma, S) \times \Sigma(\Gamma, S) = \Sigma(\Gamma, S \times S)$$

is a subshift of finite type.

(*iii*)   *There exists an integer N such that, for every $\xi \in \partial\Gamma$, $card(\pi_0^{-1}(\xi)) \leq N$.*

Properties (*i*) and (*ii*) immediately imply the following

Corollary 8.2. — *The dynamical system $(\partial\Gamma, \Gamma)$ is finitely presented.*   ∎

The proof of Theorem 8.1 is divided into several lemmas. Let us begin by introducing a few notations.

From now on, $X = K(\Gamma, G)$ is the Cayley graph of Γ relatively to a finite generating set $G \subset \Gamma$. Equipped with its canonical metric, X is a geodesic space which is proper. Let δ be a real number such that X is δ-hyperbolic. As a graph, X has bounded geometry, and therefore satisfies the hypotheses made in §7. Let Φ_0 be the set of integral cocycles which were defined in §7. The group Γ acts continuously on Φ_0 (we have $\Gamma \subset Aut(X)$ by making Γ act by left translations on on X). Let us fix an integer $d \geq 30\delta + 1$.

Let B be the closed ball in X centered at the identity element Id of Γ and of radius $3d$. We take as a set of symbols S the set of restrictions to $B \times B$ of the elements of Φ_0. By Lemma 7.7, the set S is finite. Consider the Bernoulli shift $\Sigma = \Sigma(\Gamma, S)$ and the map $P : \Phi_0 \to \Sigma$ which associates to each cocycle $\varphi \in \Phi_0$ the map $\sigma = P(\varphi) : \Gamma \to S$ defined by

$$\sigma(\gamma) = \gamma\varphi_{|B \times B} \text{ for every } \gamma \in \Gamma.$$

(The restriction of $\gamma\varphi$ to $B \times B$ belongs to S, because Φ_0 is Γ-invariant.)

Lemma 8.3. — *The map P is injective, continuous and Γ-equivariant.*

PROOF. Let φ_1 and φ_2 be elements of Φ_0 such that $P(\varphi_1) = P(\varphi_2)$. Let x and y be elements in X such that $\mid x - y \mid\leq 3d$. It is clear that there exists an element γ in Γ such that γx and $\gamma y \in B$ (we can take, for example, γ such that γ^{-1} is as close as possible from the middle of a geodesic segment $[x, y]$). Using the fact that $\gamma\varphi_1$ and $\gamma\varphi_2$ have the same restriction on $B \times B$, we deduce that

$$\varphi_1(x, y) = \gamma\varphi_1(\gamma x, \gamma y)$$

$$= \gamma\varphi_2(\gamma x, \gamma y)$$

$$= \varphi_2(x, y).$$

This implies $\varphi_1 = \varphi_2$ (recall that an element of Φ_0 is determined by its restriction to the set of $(x, y) \in X \times X$ such that $\mid x - y \mid\leq 3d$). This proves the injectivity of P.

Let (φ_n) be now a sequence of elements of Φ_0 which converges to φ. Consider an element γ of Γ. As the topology of Φ_0 is the topology of uniform convergence on compact sets in $X \times X$, and as the elements of Φ_0 take integer values on $X^0 \times X^0$, there exists an integer $n(\gamma)$ such that $\gamma\varphi_n(x,y) = \gamma\varphi(x,y)$ for $(x,y) \in B \times B$ and for every $n \geq n(\gamma)$. The sequence $P(\varphi_n)$ converges therefore to $P(\varphi)$, which proves the continuity of P.

For every $\varphi \in \Phi_0$ and for every $\gamma, \gamma' \in \Gamma$, we have

$$\gamma P(\varphi)(\gamma') = P(\varphi)(\gamma'\gamma) = \gamma'\gamma\varphi_{|B\times B} = P(\gamma\varphi)(\gamma'),$$

and therefore $\gamma P(\varphi) = P(\gamma\varphi)$. This proves the Γ-equivariance of P, and finishes the proof of the lemma. ∎

Using the compactness of Φ_0, we see that P defines a homeomorphism from Φ_0 onto its image. From now on, we identify φ and $P(\varphi)$ for every $\varphi \in \Phi_0$. Hence, Φ_0 is a subset of Σ.

Lemma 8.4. — *Φ_0 is a subshift of finite type.*

PROOF. For the proof, we construct a cylinder $C \subset \Sigma$ such that $\Phi_0 = \cap_{\gamma\in\Gamma}\gamma^{-1}C$. Let $F \subset \Gamma$ be the set of $\alpha \in \Gamma$ such that $\mid \alpha \mid \leq 6d$. Note that F is a finite set. Let A be the set of maps $f : F \to S$ satisfying the following property:

(∗)   For every $\alpha \in F$ and for every $(x,y) \in B \times B$ such that $(\alpha x, \alpha y) \in B \times B$, we have

$$f(\alpha)(\alpha x, \alpha y) = f(Id)(x,y).$$

Let C be the cylinder with basis A, that is to say,

$$C = \{\sigma : \Gamma \to S \mid \sigma_{|F} \in A\}.$$

If φ is an element of Φ_0, we have, for all $\alpha \in F$ and for all $(x,y) \in B \times B$ such that $(\alpha x, \alpha y) \in B \times B$,

$$\varphi(\alpha)(\alpha x, \alpha y) = \alpha\varphi(\alpha x, \alpha y) = \varphi(x,y) = \varphi(Id)(x,y).$$

It follows that $\varphi \in C$. As Φ_0 is Γ-invariant, we conclude that $\Phi_0 \subset \cap_{\gamma\in\Gamma}\gamma^{-1}C$.

Let σ be now an element of $\cap_{\gamma\in\Gamma}\gamma^{-1}C$. For every x and y in X such that $\mid x - y \mid \leq 3d$, let us define

(8.4.1) $\psi(x,y) = \sigma(\gamma)(\gamma x, \gamma y),$

by taking $\gamma \in \Gamma$ such that $(\gamma x, \gamma y) \in B \times B$. In fact, $\psi(x,y)$ does not depend on the choice of γ. Indeed, let us take another element γ' of Γ such that $(\gamma'x, \gamma'y) \in B \times B$. Define $\alpha = \gamma\gamma'^{-1}$. Using the triangular inequality, we obtain

$$\mid \alpha \mid = \mid \gamma'^{-1} - \gamma^{-1} \mid$$

$$\leq |\, \gamma'^{-1} - x \,| + |\, x - \gamma^{-1} \,|$$

$$= |\, Id - \gamma'x \,| + |\, \gamma x - Id \,|$$

$$\leq 3d + 3d = 6d,$$

which implies that $\alpha \in F$.

Using the fact that $\gamma'\sigma \in C$, we have

$$\sigma(\gamma')(\gamma'x, \gamma'y) = \gamma'\sigma(Id)(\gamma'x, \gamma'y)$$

$$= \gamma'\sigma(\alpha)(\alpha\gamma'x, \alpha\gamma'y) \text{ according to } (*),$$

$$= \sigma(\gamma)(\gamma x, \gamma y).$$

Consequently, formula (8.4.1) defines a function

$$\psi : \{(x,y) \in X \times X \text{ such that } |\, x - y \,| \leq 3d\} \to \mathbb{R}.$$

From (8.4.1), we know that, for every fixed $x \in X$, the restriction of ψ to the set of ordered pairs (x,y) such that $|\, x - y \,| \leq 3d$ is the restriction of an element of Φ_0. The "local" nature of properties (i), (ii) and (iii) which define Φ in §1 shows that $\psi \in \Phi_0$. On the other hand, for every $\gamma \in \Gamma$ and for every $(x,y) \in B \times B$, we have

$$\sigma(\gamma)(x,y) = \psi(\gamma^{-1}x, \gamma^{-1}y)$$

$$= \psi(\gamma)(x,y).$$

Therefore, $\sigma = \psi \in \Phi_0$, which completes the proof of Lemma 8.4. ∎

Let $\pi_0 : \Phi_0 \to \partial X = \partial\Gamma$ be the map which associates to every element of Φ_0 its point at infinity, and let $R(\pi_0) \subset \Sigma \times \Sigma = \Sigma(\Gamma, S \times S)$ be the equivalence relation associated to π_0, *i.e:*

$$R(\pi_0) = \{(\varphi_1, \varphi_2) \mid \pi_0(\varphi_1) = \pi_0(\varphi_2)\}.$$

Lemma 8.5. — $R(\pi_0)$ *is a subshift of finite type of* $\Sigma \times \Sigma = \Sigma(\Gamma, S \times S)$.

PROOF. If φ_1 and φ_2 are elements in Φ_0, we have, by Proposition 6.5,

$$(8.5.1) \quad (\varphi_1, \varphi_2) \in R(\pi_0) \iff |\, \varphi_1 - \varphi_2 \,|_d \leq K,$$

with $K = 52\delta$.

Let D_0 be the subset of $S \times S$ which consists of the ordered pairs $(s_1, s_2) \in S \times S$, where $s_1, s_2 : B \times B \to \mathbb{R}$ satisfy

$$sup\{|\, s_1(x,y) - s_2(x,y) \,|\} \leq K,$$

the sup being taken on the set of $(x,y) \in B \times B$ such that $|\, x - y \,| \leq d$. Let $D \subset \Sigma \times \Sigma$ be the cylinder

$$D = \{(\varphi_1, \varphi_2) \in \Sigma \times \Sigma \mid (\varphi_1(Id), \varphi_2(Id)) \in D_0\}.$$

Consider, as in the preceding lemma, a cylinder $C \subset \Sigma$ such that $\Phi_0 = \cap_{\sigma \in \Gamma} \gamma^{-1} C$. Let L be the cylinder

$$L = D \cap (C \times C) \subset \Sigma \times \Sigma.$$

We can deduce immediately from (8.5.1) that

$$R(\pi_0) = \cap_{\gamma \in \Gamma} \gamma^{-1} L,$$

which shows that $R(\pi_0)$ is a subshift of finite type. ∎

Proof of Theorem 8.1. — Theorem 8.1 results from the preceding lemmas, together with Proposition 7.4 and Proposition 7.9. ∎

Exercise. Prove directly that for every hyperbolic group Γ, the dynamical system $(\partial \Gamma, \Gamma)$ is expansive. (Note that this allows us to recover the fact that $R(\pi_0)$ is of finite type, using Proposition 4.3 of Chapter 2).
Hint: If ξ and η are distinct points of $\partial \Gamma$, and if g is is a geodesic in X joining these two points, then there exists an element $\gamma \in \Gamma$ such that the geodesic γg passes through the identity element of Γ. We can then deduce an expansivity constant with respect to the visual metric on $\partial \Gamma$ (*cf.* Chapter 1, §4).

Here are two more corollaries of Theorem 8.1.

Corollary 8.6. — *Let Γ be a group containing a finite-index subgroup Γ' which is a free group of finite rank (note that this implies that the group Γ is hyperbolic). Then the dynamical system $(\partial \Gamma, \Gamma)$ is a sofic system.*

PROOF. We apply Proposition 6.1 of Chapter 2, using the fact that $\partial \Gamma$ is homeomorphic to $\partial \Gamma'$, and therefore is totally disconnected. ∎

Example. $SL_2(\mathbb{Z})$ contains a free group of rank two as a subgroup of index 12. The action of $SL_2(\mathbb{Z})$ on its boundary – which is a Cantor set – defines therefore a sofic system.

Corollary 8.7. — *Let X be a geodesic space which is proper and hyperbolic. Let Γ be a group of isometries of X acting cocompactly and properly discontinuously on this space. Then the action of Γ on ∂X defines a finitely presented dynamical system. Furthermore, there exists a finite presentation $\pi : \Phi \to \partial X$ of the system $(\partial X, \Gamma)$ and an integer N such that $\mathrm{card}\big(\pi^{-1}(\xi)\big) \leq N$ for every $\xi \in \partial X$.*

PROOF. By Theorem 6.5 of Chapter 1, the group Γ is hyperbolic and the dynamical system $(\partial X, \Gamma)$ is topologically conjugate to the dynamical system $(\partial \Gamma, \Gamma)$. ∎

Example. Let V be a compact simply connected Riemannian manifold of negative curvature, and let $\tilde{V}$ be the universal cover of V. Then the action of $\pi_1(V)$ on $\partial \tilde{V}$ is finitely presented.

Chapter 3. — The boundary as a finitely presented dynamical system

Notes and comments on Chapter 3

This chapter is based essentially on §7.5.E of [Gro 3]. Theorem 8.1 (finite presentation of the action of a hyperbolic group on its boundary) and its corollaries constitute an adaptation of Theorem 8.4.C of [Gro 3] (see also [Gro 1], §5.4 and §6.2).

Bibliography for Chapter 3

[BGS] W. Ballmann, M. Gromov and V. Schroeder, "Manifolds of nonpositive curvature", Progress in Mathematics, vol. **61**, Birkhäuser, (1985).

[Gro 1] M. Gromov, "Hyperbolic manifolds, groups and actions", *Ann. of Math. Studies* **97**, Princeton University Press (1982),pp.183-215.

[Gro 3] ——, "Hyperbolic groups", *in* Essays in Group Theory, MSRI publ. **8**, Springer, 1987, pp.75-263.

[Kel] J. L. Kelley, "General Topology", Springer-Verlag, 1975.

Chapter 4

Another finite presentation for the action of a hyperbolic group on its boundary

Let Γ be a hyperbolic group equipped with a finite generating set, together with the associated word-metric.

We have seen in the preceding chapter that the dynamical system $(\partial\Gamma, \Gamma)$ is finitely presented, and for that we have constructed a finite presentation whose set of symbols is a certain set of "cocycles", or "1-forms", wich are defined on a certain ball of Γ centered at the identity element.

It is interesting to have several different presentations of the same dynamical system. In this chapter, we describe a new finite presentation of the dynamical system $(\partial\Gamma, \Gamma)$. The set of symbols that is used for this presentation is a set of "relations" in Γ, which we can think of as a set of "discrete vectorfields" which are again defined on a certain ball in Γ which is centered at the identity and which satisfy certain conditions for which these fields will be called "convergent quasi-geodesic fields".

§1 – Convergent sequences

Let X be a proper geodesic space which is δ-hyperbolic. Without loss of generality, we suppose that $\delta > 0$. In this paragraph, we establish a sufficient condition, which will be useful for us in the sequel, for the convergence of a sequence of points in X to a point in ∂X. The techniques which are involved are elementary (they make use only of the definition of the δ-hyperbolicity in terms of the Gromov product).

In this paragraph, (x_n) is a sequence of points in X satisfying the following two properties:

(1.1.1) $\exists A \geq 2000\delta$ such that $\forall i = 0, 1, 2, ...,$ we have $A \leq \mid x_i - x_{i+1} \mid \leq A + 10\delta$.

(1.1.2) $\forall i = 0, 1, 2, ...,$ we have $(x_i.x_{i+2})_{x_{i+1}} \leq 10\delta$.

Lemma 1.1. — *For such a sequence (x_n), we have, for every $n \geq 0$, for every $i = 0, 1, 2, ..., n$ and for every geodesic segment $[x_0, x_n]$,*

$$dist(x_i, [x_0, x_n]) \leq 16\delta.$$

PROOF. Let us begin by proving that for every $i = 0, 1, 2, ...,$ we have

(1.1.3) $dist(x_i, [x_0, x_{i+1}]) \leq 15\delta.$

The proof is by induction. For that, let us take inequality (1.1.3) as an induction hypothesis, and let us prove that it implies:

$$dist(x_{i+1}, [x_0, x_{i+2}]) \leq 15\delta.$$

By δ-hyperbolicity, we have:

(1.1.4) $(x_i.x_{i+2})_{x_{i+1}} \geq min\big((x_0.x_{i+2})_{x_{i+1}}, (x_0.x_i)_{x_{i+1}}\big) - \delta.$

The left hand side, $(x_i.x_{i+2})_{x_{i+1}}$, is bounded above by 10δ (hypothesis 1.1.2). On the other hand, let us observe that the term $(x_0.x_i)_{x_{i+1}}$ in the right hand side is large compared to 10δ. Indeed, we have, by the induction hypothesis,

$$dist(x_i, [x_0, x_{i+1}]) \leq 15\delta.$$

Now we use the fact that $\mid x_i - x_{i+1} \mid \geq A$ (hypothesis (1.1.1.)). If the space X was a tree, this would imply (see Figure 1):

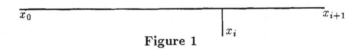

Figure 1

70

(1.1.5) $dist(x_{i+1}, [x_0, x_i]) \geq A - 15\delta.$

For a general δ-hyperbolic space X, we use the theorem of approximation by trees, applied to the segments $[x_i, x_0]$ and $[x_i, x_{i+1}]$, taking x_i as a basepoint (Chapter 1, Theorem 5.1). This gives us, instead of (1.1.5):

$$dist(x_{i+1}, [x_0, x_i]) \geq A - 10\delta - 4\delta \geq 2000\delta - 14\delta,$$

which implies (*cf.* Chapter 1, Proposition 1.5):

$$(x_0.x_i)_{x_{i+1}} \geq 2000\delta - 18\delta.$$

Therefore, we have

$$(x_0.x_i)_{x_{i+1}} - \delta > (x_i.x_{i+2})_{x_{i+1}}.$$

Inequality (1.1.4) implies then:

$$(x_i.x_{i+2})_{x_{i+1}} \geq (x_0.x_{i+2})_{x_{i+1}} - \delta,$$

and therefore

$$(x_0.x_{i+2})_{x_{i+1}} \leq 10\delta + \delta = 11\delta.$$

Again by Proposition 1.5 of Chapter 1, we conclude that:

$$dist(x_{i+1}, [x_0, x_{i+2}]) \leq 11\delta + 4\delta = 15\delta.$$

This concludes the proof of inequality (1.1.3) for every $i = 0, 1, 2, ...$
For $0 \leq i \leq n - 1$, we have, by δ-hyperbolicity:

(1.1.6) $(x_{i-1}.x_{i+1})_{x_i} \geq min((x_0.x_{i-1})_{x_i}, (x_0.x_{i+1})_{x_i}) - \delta.$

Using (1.1.3), and by the same reasoning which we did for the proof of inequality (1.1.4), we see that the term $(x_0.x_{i-1})_{x_i}$ in (1.1.6) is large compared to the left hand side of this inequality (which, by hypothesis (1.1.2), is bounded above by 10δ).
We obtain therefore:

$$(x_{i_1}.x_{i+1})_{x_i} \geq (x_0.x_{i+1})_{x_i} - \delta.$$

Using again the δ-hyperbolicity, we obtain:

$$(x_{i-1}.x_{i+1})_{x_i} \geq min((x_0.x_n)_{x_i}, (x_{i+1}.x_n)_{x_i}) - 2\delta.$$

By the same reasonig which we did above (applied to the sequence of points x_n, ..., x_{i+1}, x_i, taken in this dicreasing order with respect to the indices, instead of the sequence $x_0, ..., x_i, x_{i+1}$), the term $(x_{i+1}.x_n)_{x_i}$ is large compared to $(x_{i-1}.x_{i+1})_{x_i}$ (this last quantity being bounded above by 10δ).
We therefore obtain:

$$(x_0.x_n)_{x_i} \leq (x_{i-1}.x_{i+1})_{x_i} + 2\delta \leq 12\delta,$$

hence by Proposition 1.5 of Chapter 1,

$$dist(x_i, [x_0, x_n]) \leq 12\delta + 4\delta = 16\delta.$$

This proves Lemma 1.1. ∎

Lemma 1.2. — *For every* $i = 0, 1, 2, ...,$ *we have* $\mid x_n - x_0 \mid \geq n(A - 32\delta)$.

PROOF. We show, again by induction, that for every $i = 1, 2, ...,$ the following inequality holds:

(1.2.1) $\mid x_0 - x_i \mid \geq \mid x_0 - x_{i-1} \mid + A - 32\delta$.

The lemma will follow immediately.

For $i = 1$, ineqality (1.2.1) is a consequence of (1.1.1).

Let us then take inequality (1.2.1) as an induction hypothesis, and let us prove that we have then:

(1.2.2) $\mid x_0 - x_{i+1} \mid \geq \mid x_0 - x_i \mid + A - 32\delta$.

By property (1.1.2), we have:

(1.2.3) $\mid x_{i+1} - x_i \mid + \mid x_{i-1} - x_i \mid - \mid x_{i+1} - x_{i-1} \mid \leq 20\delta$.

On the other hand, the δ-hyperbolicity condition applied to the four points x_0, x_{i-1}, x_i and x_{i+1} (*cf.* Chapter 1, Proposition 1.7) gives:

(1.2.4) $\mid x_0 - x_i \mid + \mid x_{i-1} - x_{i+1} \mid$

$$\leq max(\mid x_0 - x_{i-1} \mid + \mid x_i - x_{i+1} \mid, \mid x_0 - x_{i+1} \mid + \mid x_i - x_{i-1} \mid) + 2\delta.$$

Suppose that, in this last inequality, the first term in the max of the right hand side is the largest term. This implies, using (1.1.1):

$$\mid x_0 - x_i \mid + \mid x_{i-1} - x_{i+1} \mid \leq \mid x_0 - x_{i-1} \mid + \mid x_i - x_{i+1} \mid + 2\delta$$

$$\leq \mid x_0 - x_{i-1} \mid + A + 10\delta + 2\delta.$$

On the other hand, we have, by (1.2.3) and (1.1.1):

$$\mid x_{i+1} - x_{i-1} \mid \geq 2A - 20\delta.$$

These last inequalities imply:

$$2A - 20\delta + \mid x_0 - x_i \mid \leq \mid x_0 - x_{i-1} \mid + A + 12\delta.$$

By the induction hypothesis (1.2.1), we obtain therefore:

$$\mid x_0 - x_{i-1} \mid + A - 32\delta + 2A - 20\delta \leq \mid x_0 - x_{i-1} \mid + A + 12\delta,$$

hence

$$2A \leq 64\delta,$$

which contradicts the fact that $A \geq 2000\delta$.

Thus, it is the second term in the max of (1.2.4) which is the largest, and we therefore have:

$$\mid x_0 - x_{i+1} \mid \geq \mid x_0 - x_i \mid + \mid x_{i-1} - x_{i+1} \mid - \mid x_i - x_{i-1} \mid - 2\delta,$$

hence

$$\mid x_0 - x_{i+1} \mid \geq \mid x_0 - x_i \mid + (2A - 20\delta) - (A + 10\delta) - 2\delta$$

$$= \mid x_0 - x_i \mid + A - 32\delta.$$

This proves Lemma 1.2. ∎

Proposition 1.3. — *The sequence (x_n) is quasi-geodesic, and therefore converges to a point $\omega \in \partial X$. Furthermore, for every geodesic γ joining the points x_0 and ω, we have, for all $i = 0, 1, 2, ...,$*

$$dist(x_i, \gamma) \leq 20\delta.$$

PROOF. For every i and j in $\mathbf{N}$, we have (using property (1.1.1) and Lemma 1.2):

$$\mid i - j \mid (A - 32\delta) \leq \mid x_i - x_j \mid \leq \mid i - j \mid (A + 10\delta),$$

which shows that the sequence (x_n) is quasi-geodesic. By Theorem 6.2 of Chapter 1, this sequence converges to a well-defined point $\omega \in \partial X$.

Let us consider now a geodesic γ joining x_0 to ω, and let us fix an integer $i \geq 0$.

For every $n \geq i$, we have, by Lemma 1.1,

$$dist(x_i, [x_0, x_n]) \leq 16\delta.$$

Let us take a sequence of points y_n on $[x_0, \omega]$ converging to ω. We have $(x_n.y_n) \to \infty$ (as the two sequences converge to the same point of ∂X), and therefore

$$(1.3.1) \quad dist(x_0, [x_n, y_n]) \to \infty.$$

For a given $i \in \mathbf{N}$, let us consider a geodesic triangle $[x_0, x_n, y_n]$. Let x_i' be a point situated on the side $[x_0, x_n]$ of this triangle, satisfying $\mid x_i - x_i' \mid \leq 16\delta$. This triangle being 4δ-narrow, we can find a point x_i'' on $[x_0, y_n] \cup [y_n, x_n]$ such that $\mid x_i' - x_i'' \mid \leq 4\delta$.

(Using (1.3.1)), we know that by taking n large enough, the point x_i'' is necessarily on the side $[x_0, y_n]$ (see Figure 2). We therefore have

$$dist(x_i, \gamma) \leq \mid x_i - x_i'' \mid \leq 20\delta.$$

∎

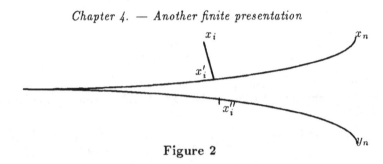

Figure 2

§2 – Convergent quasi-geodesic fields

Let $\mathcal{R}$ be a subset of $X \times X$. Then $\mathcal{R}$ defines a relation on X, and we shall write $x\mathcal{R}y$ to express that $(x, y) \in \mathcal{R}$.

Definition 2.1. — We say that the relation $\mathcal{R}$ is a *convergent quasi-geodesic field* on X if the following five conditions are satisfied:

(2.1.1) $\exists A \geq 2000\delta$ such that $x\mathcal{R}y \Rightarrow A \leq |x - y| \leq A + 10\delta$.

(2.1.2) If $x\mathcal{R}y$ and $y\mathcal{R}z$ then $(x.z)_y \leq 10\delta$.

(2.1.3) If $|x - x'| \leq 50\delta$, and if $x\mathcal{R}y$ and $x'\mathcal{R}y'$, then $|y - y'| \leq 100\delta$.

(2.1.4) If $x\mathcal{R}y$, $y\mathcal{R}z$, and $t\mathcal{R}u$, and if $(x.y)_t \leq 150\delta$, then $(t.u)_y \leq 170\delta$ and $(y.z)_u \leq 190\delta$.

(2.1.5) The projection of $\mathcal{R}$ on the first factor is surjective.

As an example, let us take an arbitrary point ζ in ∂X, and let us fix a number $A \geq 2000\delta$. We define then the following relation $\mathcal{R}_0$ on X:

$x\mathcal{R}_0 y$ if and only if y is on a geodesic ray $[x, \zeta[$ and $|x - y| = A$.

Proposition 2.2. — *The relation $\mathcal{R}_0$ is a convergent quasi-geodesic field.*

PROOF. (sketch) Properties (2.1.1) and (2.1.5) are trivially true. The three other properties can easily be checked in the case where X is a tree (with constants that are smaller than those which appear in Definition 2.1), and then, for the general case, the theorem of approximation by trees can be applied. ∎

Let $\mathcal{R}$ be a convergent quasi-geodesic field, and (x_n) a sequence of points in X such that $x_i\mathcal{R}x_{i+1}$ for every $i = 0, 1, 2, \dots$. This sequence satisfies therefore conditions (1.1.1) and (1.1.2) of §1, which, as we have already seen, imply that the sequence converges to a point $\omega \in \partial X$. The next proposition shows that this point is canonically associated to the relation $\mathcal{R}$ (that is, the point does not depend on the chosen sequence x_n).

Proposition 2.3. — *Let (x_n) and (y_n) be two sequences satisfying $x_i\mathcal{R}x_{i+1}$ and $y_i\mathcal{R}y_{i+1}$ for all $i = 0, 1, 2, \dots$. Then $\lim_{n\to\infty} x_n = \lim_{n\to\infty} y_n$.*

PROOF. Let us recall that we have supposed that $\delta > 0$. For the proof, we can assume that the points x_0 and y_0 satisfy the condition $\mid x_0 - y_0 \mid \leq \delta$ (by (2.1.5)).

Let $\omega = lim_{n\to\infty}x_n$ and $\omega' = lim_{n\to\infty}y_n$. We must show that $\omega = \omega'$.

Suppose that $\omega \neq \omega'$, and consider two geodesic rays $\gamma = [x_0, \omega[$ and $\gamma' = [y_0, \omega'[$, a geodesic segment $[x_0, y_0]$ and a geodesic $]\omega, \omega'[$. We will need the following

Lemma 2.4. — *We can find three points, a, a' and a'' respectively on γ, γ' and $]\omega, \omega'[$, such that*

$$\mid a - a'' \mid \leq 17\delta,$$

$$\mid a' - a'' \mid \leq 17\delta,$$

and therefore also

$$\mid a - a' \mid \leq 34\delta.$$

PROOF. Let p be an arbitrary point on the geodesic $]\omega, \omega'[$ and consider the function

$$dist(p, \gamma) - dist(p, \gamma').$$

When p tends to ω, this function tends to ∞, and when p tends to ω', the function tends to ∞. We can therefore find a point $p = a'' \in]\omega, \omega'[$ satisfying $dist(a'', \gamma) = dist(a'', \gamma')$.

Consider the geodesic quadrilateral $[x_0, y_0, \omega', \omega]$. By Proposition 3.2 of Chapter 1, every point on $]\omega, \omega'[$ is at distance $\leq 16\delta$ from the union $\gamma \cup [x_0, y_0] \cup \gamma'$, and is therefore at distance $\leq 17\delta$ from $\gamma \cup \gamma'$. Thus, there exists a point $a \in \gamma$ such that $\mid a - a'' \mid \leq 17\delta$ and a point $a' \in \gamma'$ such that $\mid a' - a'' \mid \leq 17\delta$. Therefore, we have also $\mid a - a' \mid \leq 34\delta$ by the triangle inequality, and Lemma 2.4 is proved. ∎

Consider now the sequences (x_i) and (y_i), and let (x_i') and (y_i') be sequences of projections of (x_i) and (y_i) respectively on γ and γ'.

Let $i \in \mathbb{N}$ be the integer such that x_i' is the last point of the sequence (x_i') satisfying $\mid x_0 - x_i' \mid \leq \mid x_0 - a \mid$, and $j \in \mathbb{N}$ the integer such that y_j' is the last point of the sequence (y_i') satisfying $\mid y_0 - y_j' \mid \leq \mid y_0 - a' \mid$ (Figure 3).

By symmetry, we can assume

(2.4.1) $\mid a - x_i' \mid \geq \mid a' - y_j' \mid$

We have then the following

Lemma 2.5. — $(x_i, x_{i+1})_{y_j} \leq 146\delta$.

PROOF. Let us show first that there exists a point $y_j'' \in [x_i', x_{i+1}']$ satisfying the following inequality:

(2.5.1) $\mid y_j' - y_j'' \mid \leq 118\delta$.

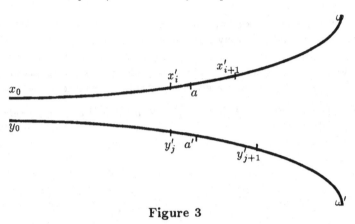

Figure 3

Let us choose a geodesic segment $[a, a']$, and let us consider the quadrilateral $[x_0, y_0, a', a]$. Every point on the side $[y_0, a']$ is at distance $\leq 8\delta$ from the union of the three other sides. Therefore, the point y'_j is at distance $\leq 34\delta + 8\delta = 42\delta$ from a point p on $[x_0, a]$.

Suppose p is on the segment $[x'_i, a]$. Inequality (2.5.1) is then satisfied with $y''_j = p$.
Suppose now that p is on $[x_0, x'_i]$. This implies that:

(2.5.2) $\quad | p - x'_i | \leq 76\delta.$

Indeed, if this were not true, we would have:

$$| a - p | > | a - x'_i | + 76\delta.$$

But we have

$$| a - p | \leq | a - a' | + | a' - y'_j | + | y'_j - p |$$

$$\leq 34\delta + | a' - y'_j | + 42\delta,$$

which would imply

$$| a - x'_i | < | a' - y'_j |,$$

contradicting inequality (2.4.1). Inequality (2.5.2) is therefore satisfied, and this implies

$$| y'_j - x'_i | \leq 42\delta + 76\delta = 118\delta.$$

Thus, in this case, inequality (2.5.1) is satisfied, with $y''_j = x'_i$.
Let us choose now geodesic segments $[x_i, x_{i+1}]$, $[x_i, x'_i]$ and $[x_{i+1}, x'_{i+1}]$, and let us consider the resulting quadrilateral $[x_i, x'_i, x'_{i+1}, x_{i+1}]$. The point y''_j on the side $[x'_i, x'_{i+1}]$ is at distance $\leq 8\delta + 20\delta = 28\delta$ from a point $y'''_j \in [x_i, x_{i+1}]$.
Hence, using (2.5.1), we have

$$| y_j - y'''_j | \leq 118\delta + 28\delta = 146\delta,$$

76

which implies
$$dist(y_j, [x_i, x_{i+1}]) \leq 146\delta.$$

Therefore, $(x_i, x_{i+1})_{y_j} \leq 146\delta$, which proves Lemma 2.5. ∎

Condition (2.1.4) implies (taking $x = x_i$, $y = x_{i+1}$, $t = y_i$ and $u = y_{i+1}$):

(2.5.3) $(y_j \cdot y_{j+1})_{x_{i+1}} \leq 170\delta.$

We can now prove the following

Lemma 2.6. — *We have $\mid a - x_{i+1} \mid \leq 334\delta$.*

PROOF. Consider a geodesic segment $[y_j, y_{j+1}]$, and let m be a projection of x_{i+1} on that segment. Inequality (2.5.3) implies:

$$\mid x_{i+1} - m \mid \leq 170\delta + 4\delta = 174\delta.$$

Let now m' be a projection of m on the segment $[y'_j, y'_{j+1}]$. By considering the geodesic quadrilateral $[y'_j, y_j, y_{j+1}, y'_{j+1}]$ and by the same kind of reasoning than the one we have used above, we conclude that

$$\mid m - m' \mid \leq 28\delta.$$

For the rest of the proof of the lemma, we distinguish two cases:

1st case. — The point m' is on the segment $[y'_j, a']$.
    The quadrilateral $[x_0, y_0, a, a']$ being 8δ-narrow, there exists a point $m'' \in [x_0, a]$, with $\mid m' - m'' \mid \leq 34\delta + 8\delta = 42\delta$.
    This implies

$$\mid x_{i+1} - m'' \mid \leq \mid x_{i+1} - m \mid + \mid m - m' \mid + \mid m' - m'' \mid$$

$$\leq 174\delta + 28\delta + 42\delta = 244\delta,$$

hence
$$\mid x'_{i+1} - m'' \mid \leq 244\delta + 20\delta = 264\delta.$$

As the three points m'', a and x'_{i+1} are aligned in this order, we have also

$$\mid x'_{i+1} - a \mid \leq 244\delta$$

and therefore
$$\mid x_{i+1} - a \mid \leq 244\delta + 20\delta = 264\delta,$$

which proves the lemma in this case.

2nd case. — m' is on the segment $[a', y'_{j+1}]$.
    By considering the geodesic triangle $[a', a'', \omega']$, we see that there exists a point m'' on $[a'', \omega'[$ satisfying $\mid m' - m'' \mid \leq 17\delta + 8\delta = 25\delta$.

In the same way, we can find a point x''_{i+1} on $[a'', \omega[$ satisfying the inequality

$$| x''_{i+1} - x'_{i+1} | \leq 25\delta.$$

Hence,

$$| x''_{i+1} - x_{i+1} | \leq 25\delta + 20\delta = 45\delta.$$

Therefore, we have:

$$| x''_{i+1} - m'' | \leq | x''_{i+1} - x_{i+1} | + | x_{i+1} - m | + | m - m' | + | m' - m'' |$$

$$\leq 45\delta + 174\delta + 28\delta + 25\delta = 272\delta,$$

and

$$| x''_{i+1} - a'' | \leq 272\delta,$$

as the three points x''_{i+1}, a'' and m'' are aligned in that order. We deduce that:

$$| x_{i+1} - a'' | \leq 272\delta + 45\delta = 317\delta,$$

and therefore

$$| x_{i+1} - a | \leq 317\delta + 17\delta = 334\delta,$$

which completes the proof of Lemma 2.6. ∎

We apply now property (2.1.4), by taking as points x, y, t, u, z in that proposition the points $x_i, x_{i+1}, y_j, y_{j+1}, x_{i+2}$ respectively.

Lemma 2.5 shows then that the hypotheses of property (2.1.4) are satisfied, and that we have therefore:

(2.6.1) $\quad (x_{i+1}.x_{i+2})_{y_{j+1}} \leq 190\delta.$

We can now prove the following

Lemma 2.7. — *We have*

$$| a' - y_{j+1} | \leq 354\delta.$$

PROOF. Consider again a geodesic segment $[x_{i+1}, x_{i+2}]$ and let m be a projection of y_{j+1} on that segment. Inequality (2.6.1) implies that

$$| y_{j+1} - m | \leq 194\delta.$$

By considering a quadrilateral $[x_{i+1}, x_{i+1}, x'_{i+2}, x'_{i+2}]$, and by a reasoning we already did above, we can find a point $m' \in [x'_{i+1}, x'_{i+2}]$ such that:

$$| y_{j+1} - m' | \leq 194\delta + 28\delta = 222\delta.$$

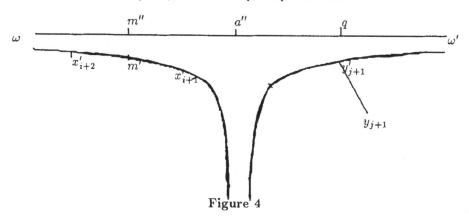

Figure 4

As y'_{j+1} is a projection of y_{j+1} on $[y_0, \omega'[$, we have

$$| y'_{j+1} - m' | \leq 222\delta + 20\delta = 242\delta.$$

Let q be now a projection of y'_{j+1} on $]\omega, \omega'[$, and m'' a projection of m' on $]\omega, \omega'[$. Suppose that the five points ω, m'', a'', q and ω' are aligned in that order (Figure 4). We shall prove the lemma under this hypothesis. The other cases can be treated in the same manner, and the lemma is true in these other cases with constants that are even smaller.

By considering the geodesic triangle $[a', a''] \cup [a'', \omega'[\cup[a', \omega'[$, we can write:

$$| y'_{j+1} - q | \leq 8\delta + 17\delta = 25\delta.$$

In the same way, by considering the triangle $[a, a''] \cup [a, \omega[\cup[a'', \omega[$, we have

$$| m' - m'' | \leq 25\delta.$$

This implies that:

$$| q - m'' | \leq | q - y'_{j+1} | + | y'_{j+1} - m' | + | m' - m'' |$$

$$\leq 25\delta + 242\delta + 25\delta = 292\delta.$$

As the points q, a'' and m' are aligned in that order, we obtain

$$| q - a'' | \leq 292\delta,$$

which gives

$$| q - a' | \leq 292\delta + 17\delta = 309\delta,$$

hence

$$| y'_{j+1} - a' | \leq | y'_{j+1} - q | + | q - a' |$$

$$\leq 25\delta + 309\delta = 334\delta,$$

and therefore

$$\mid y_{j+1} - a' \mid \leq 334\delta + 20\delta = 354\delta.$$

Lemma 2.7 is proved in this case. ∎

Lemmas 2.6 and 2.7 imply:

$$\mid x_{i+1} - y_{j+1} \mid \leq 334\delta + 354\delta$$

$$= 688\delta \leq 20 \times 50\delta.$$

By applying 20 times property (2.1.3), we obtain:

(2.7.1) $\mid x_{i+2} - y_{j+2} \mid \leq 2000\delta.$

The following lemma shows that this leads to a contradiction:

Lemma 2.8. — *We have*

$$\mid x_{i+2} - y_{j+2} \mid > 2A - 1000\delta,$$

and therefore:

$$\mid x_{i+2} - y_{j+2} \mid > 1000\delta.$$

PROOF. This will be a consequence of the fact that the four points y_{j+2}, y_{j+1}, x_{i+1} and x_{i+2} are almost aligned in that order.

Consider again projections y'_{j+1} and y'_{j+2} of the points y_{j+1} and y_{j+2} on $[y_0, \omega'[$, and projections x'_{j+1} and x'_{j+2} of the points x_{j+1} and x_{j+2} on $[x_0, \omega[$. Each of the points that we are considering is at distance $\leq 20\delta$ from its projection.

Consider also projections y^*_{j+2}, y^*_{j+1}, x^*_{i+1} and x^*_{i+2} respectively of the four points y'_{j+2}, y'_{j+1}, x'_{i+1} and x'_{i+2} on the geodesic $]\omega, \omega'[$. Each of these new four points is at distance $\leq 25\delta$ from its projection.

By (2.1.1) and the triangle inequality, we have:

$$\mid y^*_{j+2} - y^*_{j+1} \mid \geq \mid y_{j+2} - y_{j+1} \mid - \mid y_{j+2} - y_{j'+2} \mid -$$

$$\mid y_{j+1} - y'_{j+1} \mid - \mid y'_{j+2} - y^*_{j+2} \mid - \mid y'_{j+1} - y^*_{j+1} \mid$$

$$\geq A - 20\delta - 20\delta - 25\delta - 25\delta$$

$$= A - 90\delta.$$

In the same manner, we have

$$\mid x^*_{i+2} - x^*_{i+1} \mid \geq A - 90\delta.$$

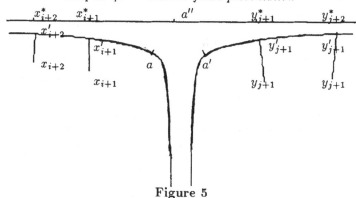

Figure 5

Suppose that the points we are considering are aligned on the geodesic $]\omega, \omega'[$ in the order indicated by Figure 5, that is, in the following order: x^*_{i+2}, x^*_{i+1}, a'', y^*_{j+1}, y^*_{j+2}. This implies that:

$$\mid x^*_{i+2} - y^*_{j+2} \mid\geq 2A - 180\delta.$$

By applying again the triangle inequality, we obtain:

$$\mid x_{i+2} - y_{j+2} \mid\geq 2A - 400\delta,$$

and Lemma 2.8 is proved.

In the case where the points on γ are aligned on $]\omega, \omega'[$ in an order which is different from that of Figure 5 (that is to say, in the following order: $x*_{i+2}$, y^*_{j+1}, x^*_{i+1}, y^*_{j+2}), we can see, by a reasoning which we have already used above, that the points x^*_{i+1} and y^*_{j+1} are close to the point a'', and therefore that the distance $\mid x^*_{i+1} - y^*_{j+1} \mid$ is not too large. We then easily see that Lemma 2.8 is also true in this case. ∎

Lemma 2.8 contradicts inequality (2.7.1), and this shows that we have necessarily $\omega = \omega'$, which concludes the proof of Proposition 2.3. ∎

Definition 2.9. — Let $\omega = lim_{n\to\infty} x_n$ be as in the preceding proposition. We say that ω is the *point at infinity associated to the field* $\mathcal{R}$, and we write $\omega = a(\mathcal{R})$.

Let now A be a real number which is $\geq 2000\delta$, and which is fixed once and for all. We denote by $\mathcal{C} = \mathcal{C}(A)$ the set of convergent quasi-geodesic fields with associated constant A (*cf.* property (2.1.1)). We take on this set $\mathcal{C}$ the topology which is induced by a topology on the set of relations on X, and which is defined as follows:

Given two relations $\mathcal{R}_1$ and $\mathcal{R}_2 \subset X \times X$, and a compact set $K \subset X \times X$, we say that $\mathcal{R}_1$ *is ϵ-close to* $\mathcal{R}_2$ *on* K if the following property is satisfied:

$$\forall (x_2, y_2) \in \mathcal{R}_2 \cap K, \exists (x_1, y_1) \in \mathcal{R}_1$$
$$\text{such that } \mid x_1 - x_2 \mid\leq \epsilon \text{ and } \mid y_1 - y_2 \mid\leq \epsilon.$$

81

Given K and ϵ as above, let $V(K, \epsilon)$ be the set of ordered pairs $(\mathcal{R}_1, \mathcal{R}_2)$ such that $\mathcal{R}_2$ is ϵ-close to $\mathcal{R}_1$ on K and $\mathcal{R}_1$ is ϵ-close to $\mathcal{R}_2$ on K.

The set of all the $V(K, \epsilon)$ (where K varies over all compact sets of $X \times X$ and ϵ varies over the positive reals) constitutes a fundamental system of neighborhoods for a uniform structure on the set of relations on X (in the sense of [Bou], §1). This structure defines a topology for the set of relations on X, and the set $\mathcal{C}$ of convergent quasi-geodesic fields associated to the constant A inherits an induced topology. We can express the convergence of a sequence of fields $\mathcal{R}_i \in \mathcal{C}$ to a field $\mathcal{R}$ in the following manner:

For every compact set $K \subset X \times X$ and for every $\epsilon > 0$, there exists an integer $i_0 \geq 0$ such that for all $i \geq i_0$, $\mathcal{R}_i$ is ϵ-close to $\mathcal{R}$ on V. By a proof which is of the same type as that of Ascoli's theorem, we can show that the space $\mathcal{C}$ is compact.

We define now the map $a : \mathcal{C} \to \partial X$ as the map which associates to each field $\mathcal{R}$ its point at infinity $a(\mathcal{R})$.

There is a left action of the group $Isom(X)$ of isometries of X on the space $\mathcal{C}$, which is defined by the following formula:

$$(x, y) \in \gamma \mathcal{R} \iff (\gamma^{-1} x, \gamma^{-1} y) \in \mathcal{R},$$

for every $\gamma \in Isom(X)$ and $\mathcal{R} \in \mathcal{C}$. We verify immediately that $\gamma \mathcal{R}$ is, as $\mathcal{R}$, a convergent quasi-geodesic field in $\mathcal{C}$.

Proposition 2.10. — *The map $a : \mathcal{C} \to \partial X$ which associates to a convergent quasi-geodesic field $\mathcal{R}$ its point at infinity $a(\mathcal{R})$ is continuous, surjective and $Isom(X)$-equivariant.*

PROOF. For the surjectivity, we consider, for every point $\zeta \in \partial X$, the convergent quasi-geodesic field $\mathcal{R}_0$ which is associated to it by Proposition 2.2. By construction, we have $a(\mathcal{R}_0) = \zeta$.

For the continuity, let $\mathcal{R}_n$ be a sequence of elements of $\mathcal{C}$ which converges to $\mathcal{R} \in \mathcal{C}$. Let $(x_n)_{n \geq 0}$ be a sequence of points of X which satisfies $x_n \mathcal{R} x_{n+1}$ for all n, and for every $i \geq 0$, let $(x_n^i)_{n \geq 0}$ be a sequence of points satisfying $x_n^i \mathcal{R}_i x_{n+1}^i$ for all n. We choose these sequences in such a way that $\forall i = 0, 1, 2, ...,$ we have $x_0 = x_0^i$.

Given $L > 0$, let $B_L(x_0)$ be the open ball centered at x_0 and of radius L. For every $\epsilon > 0$ and for every i which is large enough, the field $\mathcal{R}_i$ is ϵ-close to the field $\mathcal{R}$ on this ball. Thus, for every $n_0 \geq 0$, we can find an i_0 such that

$$\mid x_n^i - x_n \mid \leq \epsilon \; \forall i \geq i_0 \text{ and } \forall n \leq n_0.$$

Let now $\omega = a(\mathcal{R})$ and $\omega_i = a(\mathcal{R}_i)$, and let $\gamma_i : [0, \infty[\to X$ be a geodesic ray and $\gamma_i : [0, \infty[\to X$ a sequence of geodesic rays with $\gamma(\infty) = \omega$ and $\gamma_i(\infty) = \omega_i$. For every i which is large enough, and for every $L > 0$, we have $\mid \gamma(t) - \gamma_i(t) \mid \leq 40\delta$, for all $0 \leq t \leq L$. This implies $(\omega.\omega_i)_{x_0} \to \infty$, and therefore $\omega_i \to \omega$.

For the equivariance, let $\mathcal{R}$ be a convergentquasi-geodesic field, and let γ be an element of $Isom(X)$. Let x_i be a sequence associated to $\mathcal{R}$ as in 2.9, satisfying $a(\mathcal{R}) = lim_{i \to \infty} x_i$. Consider the sequence (γx_i). By the definition of the field $\gamma \mathcal{R}$, this sequence satisfies

$$lim_{i \to \infty}(\gamma x_i) = a(\gamma \mathcal{R}).$$

Thus, we have $\gamma a(\mathcal{R}) = a(\gamma \mathcal{R})$ and the map a is $Isom(X)$-equivariant. ∎

§3 – Integral fields on a graph

In what follows, the space X is a connected locally finite graph, which is δ-hyperbolic for its canonical metric (defined in Chapter 3, §7), and M is an integer such that for every vertex s of X, the number of edges that contain this vertex is bounded above by M. Let X^0 be the set of vertices of X. We suppose without loss of generality that the constant A (whose definition is the same as in the preceding sections) is an integer, and we denote by $\mathcal{C}_0 = \mathcal{C}_0(A)$ the set of convergent quasi-geodesic fields on X, with associated constant A, and which are contained in the set of vertices $X^0 \times X^0$ of $X \times X$. Of course, for the definition of the fields in $\mathcal{C}_0$, property (2.5.1) of convergent quasi-geodesic fields is replaced by the following property:

$(2.5.1)'$ The projection of $\mathcal{R}$ on the first factor of $X^0 \times X^0$ is surjective.

We shall say that an element in $\mathcal{C}_0$ is an *integral convergent quasi-geodesic field* (or, to abbreviate, an *integral field*).

Let us consider the set of relations in X^0, and let us define a topology on this set, in the same manner in which we defined a topology on the set of relations in the space X. We obtain an induced topology on the set $\mathcal{C}_0$ which makes this space compact, and convergence in $\mathcal{C}_0$ can be described in the following manner:

If $(\mathcal{R}_i)_{i \geq 0}$ is a sequence of integral fields, then $\mathcal{R}_i$ converges to an integral field $\mathcal{R} \in \mathcal{C}_0$ if and only if the following condition is satisfied::

For every finite subset of vertices $K^0 \subset X^0 \times X^0$, there is an integer $i_0 \geq 0$ such that for all $i \geq i_0$, the restriction of $\mathcal{R}_i$ on K^0 coincides with the restriction of $\mathcal{R}$ on this set.

In the same way as for the space $\mathcal{C}$, we have a map from $\mathcal{C}_0$ to ∂X, which we denote by a_0, which associates to every integral field its point at infinity.

Proposition 3.1. — *The map $a_0 : \mathcal{C}_0 \to \partial X$ is continuous, surjective and $Aut(X)$-equivariant.*

PROOF. Using Proposition 2.10, all what remains to show is surjectivity. For that, it is sufficient to consider fields of Proposition 2.2, after modifying the definition in the following way:

$\mathcal{R}_0 \subset X^0 \times X^0$ and

$x\mathcal{R}_0 y$ if and only if y is on a geodesic ray $[x,\zeta]$ and $\mid x-y\mid = A$.

Using the fact that the constant A is an integer, we see easily that $\mathcal{R}_0$ is an integral convergent quasi-geodesic field. ∎

Proposition 3.2. — *Let $\mathcal{R}_1$ and $\mathcal{R}_2 \in C_0$. We have $a_0(\mathcal{R}_1) = a_0(\mathcal{R}_2)$ if and only if the following condition is satisfied:*

$$\forall x, y_1, y_2 \in X^0, \text{ if } x\mathcal{R}_1 y_1 \text{ and } x\mathcal{R}_2 y_2, \text{ then } \mid y_1 - y_2 \mid \le 80\delta.$$

PROOF. Suppose that $\mathcal{R}_1$ and $\mathcal{R}_2$ have the same point at infinity, ζ , and let x, y_1, y_2 be three points in X^0 with $x\mathcal{R}_1 y_1$ and $x\mathcal{R}_2 y_2$. This implies that:

$$A \le \mid x-y_1 \mid \le A + 10\delta$$

and

$$A \le \mid x-y_2 \mid \le A + 10\delta.$$

Let γ be a geodesic ray between x and ζ. By Proposition 1.3, we have $dist(y_1, \gamma) \le 20\delta$ and $dist(y_2, \gamma) \le 20\delta$. If the space X were a tree, we would have immediately (see Figure 6) $\mid y_1 - y_2 \mid \le 20\delta + 20\delta + 10\delta + 10\delta = 60\delta$. In the case of a general δ-hyperbolic space, we use the theorem of approximation by trees (applied to the four points x, y_1, y_2 and ζ), and we obtain the desired inequality.

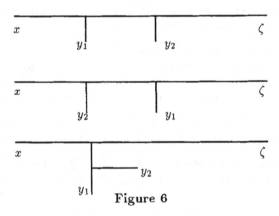

Figure 6

Conversely, suppose that $a_0(\mathcal{R}_1) \ne a_0(\mathcal{R}_2)$ and let γ be a geodesic joining these two points of ∂X. Let us take $x \in \gamma$, and let y_1 and y_2 be two points in X satisfying $x\mathcal{R}_1 y_1$ and $x\mathcal{R}_2 y_2$. Let also γ_1 and γ_2 be the two geodesic rays contained in γ, whose origin is x and which converge respectively to $a_0(\mathcal{R}_1)$ and $a_0(\mathcal{R}_2)$ (Figure 7).

Let y_1' and y_2' be projections of y_1 and y_2 on γ. We have $y_1' \in \gamma_1$ and $y_2' \in \gamma_2$, $\mid x - y_1' \mid \ge 2000\delta - 20\delta$ and $\mid x - y_2' \mid \ge 2000\delta - 20\delta$. As the points x, y_1' and y_2' are on

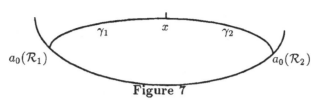

Figure 7

the same geodesic, we deduce that $\mid y_1' - y_2' \mid \geq 2000\delta - 40\delta$, which implies in particular $\mid y_1 - y_2 \mid \geq 80\delta$. This finishes the proof of the proposition. ∎

§4 – Another finite presentation of the boundary of a hyperbolic group

Let Γ be a hyperbolic group. In this paragraph, we give another proof of the following theorem (which should be compared with Theorem 8.1 of Chapter 3):

Theorem 4.1. — *There exists a finite set S, a subshift of finite type $\Phi_0 \subset \Sigma(\Gamma, S)$ and a map $\pi_0 : \Phi_0 \to \partial\Gamma$ which has the following two properties:*

(i) *π_0 is continuous, surjective and Γ-equivariant.*

(ii) *The equivalence relation*

$$R(\pi_0) = \{(\varphi_1, \varphi_2) \mid \pi_0(\varphi_1) = \pi_0(\varphi_2)\} \subset \Sigma(\Gamma, S) \times \Sigma(\Gamma, S) = \Sigma(\Gamma, S \times S)$$

is a subshift of finite type.

Before giving the proof, let us note, as we did for Theorem 8.1 of Chapter 3, the following corollary, which is merely another formulation of the theorem.

Corollary 4.2. — *The dynamical system $(\partial\Gamma, \Gamma)$ is finitely presented.* ∎

Remark. Theorem 8.1 of Chapter 3 gives more information than the theorem which we present here, since the map π_0 which we describe here is not finite-to-one.

For the proof of Theorem 4.1, the outline is the same as for the proof of Theorem 8.1 of Chapter 3. Let G be a finite set of generators for Γ, and let X be the Cayley graph associated to (Γ, G), equipped with its canonical metric. We suppose that X is δ-hyperbolic. Let A be an integer which is $\geq 2000\delta$, and let $\mathcal{C}_0$ be the associated space of integral convergent quasi-geodesic fields, as defined in §3 above. The group Γ has a left action on X and, as a subgroup of the group of simplicial automorphisms of X, it acts on $\mathcal{C}_0$.

Let B be the closed ball in X which is centered at the identity element of Γ and whose radius is $3(A + 10\delta)$, and let B^0 be the set of vertices of B.

The set S is defined as the set of restrictions to $B^0 \times B^0$ of an integral field $\mathcal{R} \in \mathcal{C}_0$. This set is, of course, finite. It will be the set of symbols for our finite presentation of $\partial X = \partial\Gamma$.

Consider the Bernoulli shift $\Sigma = \Sigma(\Gamma, S)$ and let P be the map from C_0 to Σ which associates to an arbitrary element $\mathcal{R} \in C_0$ the map $\sigma = P(\mathcal{R}) : \Gamma \to S$ defined by the formula:

$$\sigma(\gamma) = \gamma \mathcal{R}_{|B^0 \times B^0} = (\gamma \mathcal{R}) \cap (B^0 \times B^0) \; \forall \gamma \in \Gamma.$$

We have the following

Proposition 4.3. — *The map P is injective, continuous and Γ-equivariant.*

PROOF. Let $\mathcal{R}_1$ and $\mathcal{R}_2$ be two elements of C_0 such that $P(\mathcal{R}_1) = P(\mathcal{R}_2)$, and let x and y be two points in X^0 such that $\mid x - y \mid \leq A + 10\delta$. We can find then an element $\gamma \in \Gamma$ such that γx and $\gamma y \in B^0$. As $\gamma \mathcal{R}_1$ and $\gamma \mathcal{R}_2$ have the same restriction to $B^0 \times B^0$, we have

$$(\gamma x, \gamma y) \in \gamma \mathcal{R}_1 \iff (\gamma x, \gamma y) \in \gamma \mathcal{R}_2.$$

For $i = 1, 2$, we have

$$(\gamma x, \gamma y) \in \gamma \mathcal{R}_i \iff (x, y) \in \mathcal{R}_i.$$

Thus,

$$(x, y) \in \mathcal{R}_1 \iff (x, y) \in \mathcal{R}_2,$$

and the two relations $\mathcal{R}_1$ and $\mathcal{R}_2$ coincide for every $x, y \in X^0$ with $\mid x - y \mid \leq A + 10\delta$, which implies that $\mathcal{R}_1 = \mathcal{R}_2$. Therefore, P is injective.

For the continuity, let $\mathcal{R}_n$ be a sequence in C_0 which converges to the element $\mathcal{R}$ of this space. By definition of the topology on C_0, there exists then, for every $\gamma \in \Gamma$, an integer $n(\gamma)$ such that for every $n \geq n(\gamma)$, the restriction of $\gamma \mathcal{R}_n$ to $B^0 \times B^0$ coincides with the restriction of $\gamma \mathcal{R}$ to $B^0 \times B^0$. Hence, $P(\mathcal{R}_n)$ converges to $P(\mathcal{R})$, and the map is continuous.

Finally, we have, from the definitions, for every $\mathcal{R} \in C_0$ and for every $\gamma, \gamma' \in \Gamma$,

$$(\gamma P(\mathcal{R}))(\gamma') = P(\mathcal{R})(\gamma'\gamma)$$

$$= \gamma'\gamma \mathcal{R}_{|B^0 \times B^0} = \gamma'(\gamma \mathcal{R}_{|B^0 \times B^0})$$

$$= P(\gamma \mathcal{R})(\gamma').$$

This proves the equivariance. ∎

The map P defines a homomorphism from C_0 to its image.

Let $\Phi_0 = P(C_0)$.

Lemma 4.4. — Φ_0 *is a subshift of finite type of* $\Sigma = \Sigma(\Gamma, S)$.

PROOF. We construct a cylinder $C \subset \Sigma$ such that $\Phi_0 = \cap_{\gamma \in \Gamma} \gamma^{-1} C$.
Let $F \subset \Gamma$ and $H \subset S^F$ be the sets defined as

$$F = \{\alpha \in \Gamma \text{ such that } \mid \alpha \mid \leq 6(A + 10\delta)\},$$

and
$$H = \{f : F \to S \text{ satisfying condition } (*)\}$$
where condition $(*)$ is the following:

$(*)$ $\quad \forall \alpha \in F, \forall (x,y) \in B^0 \times B^0$ such that $(\alpha x, \alpha y) \in B^0 \times B^0$,

we have $(\alpha x, \alpha y) \in f(\alpha) \iff (x,y) \in f(Id)$.

Let C be the cylinder of basis H, *i.e.*

$$C = \{\sigma : \Gamma \to S | \sigma_{|F} \in H\}.$$

Let φ be an arbitrary element of Φ_0. Associated to φ, there is an element $\mathcal{R} \in \mathcal{C}_0$ such that $\varphi(\gamma) = \gamma \mathcal{R}_{|B^0 \times B^0}$.

Let now $\alpha \in F$, and (x,y) an element of $B^0 \times B^0$ such that $(\alpha x, \alpha y) \in B^0 \times B^0$. We have

$$(\alpha x, \alpha y) \in \varphi(\alpha) \iff (\alpha x, \alpha y) \in \alpha \mathcal{R}$$

$$\iff (x,y) \in \mathcal{R} \iff (x,y) \in \varphi(Id).$$

Therefore, $\varphi \in C$, and $\Phi_0 \subset C$. As Φ_0 is Γ-invariant, we have

$$\Phi_0 \subset \cap_{\gamma \in \Gamma} \gamma^{-1} C.$$

Conversely, let $\varphi \in \cap_{\gamma \in \Gamma} \gamma^{-1} C$. Let $\mathcal{R}$ be the relation in X^0 defined by:

$$x \mathcal{R} y \iff |x - y| \le 3(A + 10\delta) \text{ and } (\gamma x, \gamma y) \in \varphi(\gamma),$$

where γ is an arbitrary element in Γ such that $(\gamma x, \gamma y) \in B^0 \times B^0$.

The definition of the relation $\mathcal{R}$ does not depend on the choice of γ. Indeed, let γ' be another element of Γ such that $(\gamma' x, \gamma' y) \in B^0 \times B^0$, and let $\alpha = \gamma \gamma'^{-1}$. We have:

$$|\alpha| = |\gamma'^{-1} - \gamma^{-1}| \le |\gamma'^{-1} - x| + |x - \gamma^{-1}|$$

$$= |Id - \gamma' x| + |\gamma x - Id|$$

$$\le 6(A + 10\delta).$$

Hence, $\alpha \in F$.

Using now the fact that $\gamma' \varphi \in C$, we have:

$$(\gamma' x, \gamma' y) \in \varphi(\gamma') \iff (\gamma' x, \gamma' y) \in \gamma' \varphi(Id)$$

$$\iff (\alpha \gamma' x, \alpha \gamma' y) \in \gamma' \varphi(\alpha) \text{ (because } \alpha \in F)$$

$$\iff (\gamma x, \gamma y) \in \varphi(\gamma).$$

The relation $\mathcal{R} \subset \{(x,y) | \ |x - y| \le 3(A + 10\delta)\} \subset X^0 \times X^0$ is therefore well-defined.

The restriction of $\mathcal{R}$ to $B^0 \times B^0$ is of course the restriction of an element of $\mathcal{C}_0$, and the local nature of properties (2.1.1) to (2.1.5) shows that $\mathcal{R} \in \Phi_0$.

Finally, for all $\gamma \in \Gamma$ and for all $(x, y) \in B^0 \times B^0$, we have:

$$(x, y) \in \varphi(\gamma) \iff (\gamma^{-1}x, \gamma^{-1}y) \in \mathcal{R}$$

$$\iff (x, y) \in P(\mathcal{R})(\gamma),$$

hence $\varphi = P(\mathcal{R}) \in \Phi_0$. ∎

Let now $\pi_0 : \Phi_0 \to \partial X$ be the map which to every element φ of Φ_0 associates its point at infinity (or, more exactly, the point at infinity of $P^{-1}(\varphi) \in \mathcal{C}_0$). Let $\mathcal{R}(\pi_0)$ be the *kernel* of π_0, that is, the set

$$\{(\varphi_1, \varphi_2) \in \Sigma \times \Sigma \mid \pi_0(\varphi_1) = \pi_0(\varphi_2)\}.$$

$\mathcal{R}(\pi_0)$ is a subshift of the Bernoulli shift $\Sigma(\Gamma, S \times S) = \Sigma \times \Sigma$.

Proposition 4.5. — *The subshift $\mathcal{R}(\pi_0)$ is of finite type.*

PROOF. Let φ_1 and φ_2 be two elements of Φ_0, represented by the relations $\mathcal{R}_1$ and $\mathcal{R}_2$. We have (Proposition 4.2):

(4.5.1) $\pi_0(\varphi_1) = \pi_0(\varphi_2)$ if and only if the following is satisfied:

$$\forall x, y_1, y_2 \in X^0 \text{ with } x\mathcal{R}_1 y_1 \text{ and } x\mathcal{R}_2 y_2, \text{ we have:}$$

$$\mid y_1 - y_2 \mid \le 80\delta.$$

Let D_0 be the set of ordered pairs $(s_1, s_2) \in S \times S$ where s_1 and s_2 are restrictions on $B^0 \times B^0$ of relations which are 80δ-close from each other on this set, in the sense defined at the beginning of §3, and let D be the cylinder defined as

$$D = \{(\varphi_1, \varphi_2) \in \Sigma \times \Sigma \mid (\varphi_1(Id), \varphi_2(Id)) \in D_0\}$$

and $C \subset \Sigma$ a cylinder such that $\Phi_0 = \cap_{\sigma \in \Gamma} \gamma^{-1} C$. If L is the cylinder

$$L = D \cap (C \times C) \subset \Sigma \times \Sigma,$$

we have, using (4.5.1),

$$\mathcal{R}(\pi_0) = \cap_{\gamma \in \Gamma} \gamma^{-1} L,$$

which shows that the subshift $\mathcal{R}(\pi_0)$ is of finite type. ∎

Proof of Theorem 4.1. — The theorem is a consequence of the preceding lemmas, combined with Proposition 3.1 ∎

Notes and comments on Chapter 4

In §7.5.H of [Gro 3], Gromov gives some indications for a "coding" of ∂X by a set of "quasi-gradient fields". The conditions that Gromov gives cannot be used as such to give a finite presentation of ∂X, and in this chapter we have given a modified version of these conditions. On the other hand, Gromov suggests the existence of a coding which is invariant by quasi-isometries, and this property is not verified by the coding that we give here (condition (2.1.1) for convergent quasi-geodesic fields is not invariant by quasi-isometries). The property is also not verified by the coding which Gromov proposes.

Bibliography for Chapter 4

[Bou] N. Bourbaki, Topologie générale, Hermann, Paris.

[Gro 1] M. Gromov, "Hyperbolic manifolds, groups and actions", *Ann. of Math. Studies* **97**, Princeton University Press, 1982 , pp. 75-263.

[Gro 3] ———, "Hyperbolic groups", *in* Essays in Group Theory, MSRI publ. **8**, Springer Verlag, 1987, pp. 236-238.

Chapter 5

Trees and hyperbolic boundary

The general idea developed in this chapter is to describe several ways of approaching the boundary of a hyperbolic space (which will be here a locally finite simplicial graph) by using trees.

Let X be a connected and locally finite simplicial graph, which is δ-hyperbolic for its canonical metric, and let x_0 be a basepoint of X. We associate to X two simplicial trees, $T_{geo}(X)$ and $T_{part}(X)$. We study the properties of these two trees and we define a surjective and continuous map from the boundary of each of them to the boundary of X. In the case of $T_{part}(X)$ and if X has bounded geometry, we shall see that the map is finite-to-one (which means that the cardinality of the fibres is bounded above by a uniform constant). The properties of these trees will be used in the sequel of these notes.

If X is the Cayley graph of a hyperbolic group equipped with a finite set of generators, we construct a third tree, $T_{lex}(X)$ and a map from $\partial T_{lex}(X)$ to ∂X which is surjective and continuous (as in the case of the preceding trees) and which, as in the case of T_{part}, is finite-to-one.

§1 – Trees and projective sequences of sets

It will be convenient for us to use the description of simplicial trees in terms of connected projective sequences of sets, a description which is contained in the book of J-P. Serre, *Arbres, amalgames, SL_2* (*cf.* [Ser], §2.2).

A *connected projective sequence of sets* is a set-theoretic diagram of the form

$$(*) \quad \ldots \to E_{i+1} \to E_i \to \ldots \to E_1 \to E_0,$$

where the set E_0 is reduced to a point.

A *pointed simplicial tree* is a simplicial tree which is equipped with a basepoint at one of its vertices.

There is a natural one-to-one correspondence between the class of pointed simplicial trees and that of connected projective sequences of sets. This correspondence is given as follows:

Let (T, x_0) be a simplicial pointed tree equipped with its canonical metric. For every $i = 0, 1, 2, \ldots$, define E_i to be the set of vertices of T whose distance to the basepoint is equal to i, that is,

$$T = \{x \in T, \mid x \mid = i\}.$$

Define then the map $E_{i+1} \to E_i$ which associates to an arbitrary point $x \in E_{i+1}$ the unique point in E_i which is situated on the geodesic joining x_0 to x. This defines our connected projective sequence of sets.

Conversely, given the connected projective sequence of sets $(*)$, we associate to it a pointed tree (T, x_0) in the following manner: The set of vertices of T is the disjoint union of the E_i, and two such vertices x and y are related by an edge if and only if one is the image of the other by one of the maps of the diagram $(*)$ above. It is clear that this defines a simplicial complex of dimension 1. We take the unique element in E_0 to be the basepoint of this complex. It is easy to see that this complex is a tree. Indeed, using the one-parameter family of maps which is defined by following the path which uses the sequences of projections $E_{i+1} \to E_i$, we define a retraction of T on its basepoint x_0. The boundary of the tree, ∂T, is naturally identified to the projective limit of the sequence $(*)$.

Remarks.
1) In the preceding definition, the adjective *connected* refers to the condition which states that the set E_0 is reduced to a point, and this insures the connectedness of the associated complex T. Without this hypothesis on E_0, the complex T would be, in general, a forest (*i.e.* a disjoint union of $card(E_0)$ trees).
2) The tree associated to the sequence $(*)$ is locally finite if and only if every set E_i is finite.

§2 – The tree $T_{\mathbf{geo}}(X)$

Let X be a connected and locally finite simplicial graph which is δ−hyperbolic, equipped with a basepoint x_0 which is a vertex of this graph.

We associate to X a tree $T_{geo}(X)$, by defining the associated connected projective sequence of sets,

$$... \to E_{i+1} \to E_i \to ... \to E_1 \to E_0,$$

in the following manner:

We take for E_i the set of geodesic segments $r : [0,i] \to X$ starting at x_0. The map $E_{i+1} \to E_i$ associates to every element of E_{i+1} its restriction to $[0,i]$.

This defines the connected projective sequence of sets, and $T_{geo}(X)$ is the associated pointed tree.

There exists a natural map from the set of vertices of $T_{geo}(X)$ to X; this is the map which associates to every vertex of $T_{geo}(X)$ the endpoint of the corresponding geodesic in X. This map extends linearly to a map $f'_{geo} : T_{geo}(X) \to X$ which sends isometrically each geodesic ray of the tree which starts at the basepoint on a geodesic ray of X which start at x_0. The set $\partial T_{geo}(X)$ is identified with the set of geodesic rays in X which start at x_0, and the map f'_{geo} induces in a natural way a map $f_{geo} : \partial T_{geo}(X) \to \partial X$, which to each geodesic ray associates its point at infinity.

To study the map f_{geo}, the following lemma will be useful to us:

Lemma 2.1. — *Let r and $r' : [0, \infty[\to X$ be two geodesic rays starting at x_0, and let $\zeta = r(\infty)$ and $\zeta' = r'(\infty)$ be their two points at infinity. Let us suppose that the rays r and r' coincide on an initial segment of length t. Then, for every geodesic γ joining ζ to ζ', we have*

$$dist(x_0, \gamma) \geq t - 12\delta.$$

PROOF. Let a be the point on r (or on r') which is situated at distance t from the basepoint (Figure 1), and consider the geodesic triangle $[a, \zeta[\cup[a, \zeta'[\cup\gamma.$

The geodesic γ is contained in the 12δ-neighborhood of the union $[a, \zeta[\cup[a, \zeta'[$ (*cf.* Chapter 1, Proposition 3.2).

Let x be a projection of x_0 on γ, that is, a point on γ which satisfies $| x_0 - x | = dist(x_0, \gamma)$, and let y be a point in $[a, \zeta[\cup[a, \zeta'[$ satisfying $| x - y | \leq 12\delta$.

By the triangle inequality, we have

$$| x_0 - y | \leq | x_0 - x | + | x - y |,$$

and therefore

$$| x_0 - x | \geq t - 12\delta,$$

93

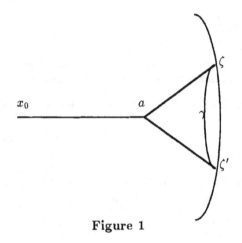

Figure 1

using the fact that

$$| \, x_0 - y \, | \geq | \, x_0 - a \, | = t \text{ and } | \, x - y \, | \leq 12\delta.$$

■

The metric spaces $T_{geo}(X)$ and X are proper (this is because, as simplicial graphs, they are locally finite), and on each of the spaces $\partial T_{geo}(X)$ and ∂X, we have a class of metrics, each metric depending on a real parameter $a > 1$ and denoted by $| \, |_a$ (these are the visual metrics which were defined in Chapter 1, §4). We recall that for the metric on ∂X, the constant a has to be in an interval $]1, a_0(\delta)[$, where $a_0(\delta)$ is a constant which depends only on δ. We shall make use of property $(P2)$ of §4 of Chapter 1, which we state here for convenience, in the form of a lemma:

Lemma 2.2. — *Let ζ and ζ' be two points in ∂X. We have:*

$$\lambda^{-1} a^{-d} \leq | \, \zeta - \zeta' \, |_a \leq \lambda a^{-d},$$

where λ is a constant which depends only on δ and on a, and where $d = dist(x_0, \gamma)$, with γ being a geodesic in X joining the points ζ and ζ'.

Let us fix now the constant $a \in]1, a_0(\delta)[$, and let us equip $\partial T_{geo}(X)$ and ∂X with their metrics $| \, |_a$. We have the following

Proposition 2.3. — *The map $f_{geo} : \partial T_{geo}(X) \to \partial X$ is Lipschitz (and therefore continuous) and surjective.*

PROOF. Let ξ and ξ' be two points of $\partial T_{geo}(X)$, and let ζ and ζ' be their images in ∂X. The two geodesic rays in $T_{geo}(X)$ starting at the basepoint and converging to ξ and ξ' coincide on an initial path of length t such that $| \, \xi - \xi' \, |_a = (2/Log \, a)a^{-t}$ (Proposition 4.1 of Chapter 1). The same thing is true for the images of these geodesics in X.

94

Let γ be a geodesic in X joining the points ζ and ζ'. Then, by Lemma 2.1, we have:

$$d = dist(x_0, \gamma) \geq t - 12\delta.$$

Lemma 2.2 implies then:

$$\mid \zeta - \zeta' \mid_a \leq \lambda a^{-(t-12\delta)} = C \mid \xi - \xi' \mid_a,$$

where C is the constant $1/2(Log\ a)a^{12\delta}\lambda$, which, as the constant λ, depends only on δ and on a. Thus, the map f_{geo} is Lipschitz.

For the surjectivity, let ζ be an arbitrary point of ∂X, and $r : [0, \infty[\to X$ a geodesic ray converging to ζ. This ray defines, in a unique way, a geodesic ray in the tree $T_{geo}(X)$, whose point at infinity is sent on ζ by the map f_{geo}. This completes the proof of the proposition. ∎

Contrarily to the fibers of the maps f_{part} and f_{flex} which will be defined in the next sections, the fibers of the map f_{geo} are in general infinite, as the following example will show:

Example. If X is the infinite graph represented in Figure 2, then ∂X has only one element, whereas $T_{geo}(X)$ is a tree whose boundary is homeomorphic to $\mathbb{Z} \cup \{-\infty, \infty\}$ (see Figure 2).

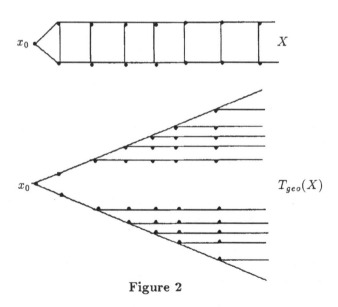

Figure 2

In Figure 3, we give an example where ∂X is again reduced to a single point and where ∂T_{geo} is a Cantor set.

95

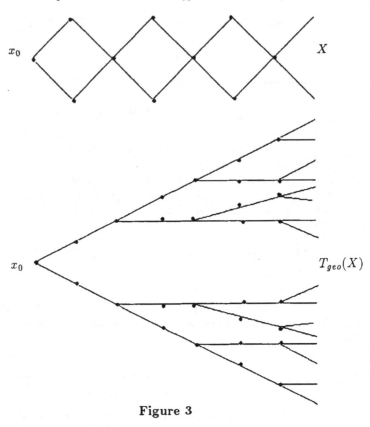

Figure 3

§3 – The tree $T_{\text{part}}(X)$

Let X be a connected simplicial graph which is locally finite and δ-hyperbolic, and let us take as a basepoint a vertex x_0 of X. Before giving the definition of the tree $T_{part}(X)$ which is associated to X, we fix a few notations and we give some preliminary lemmas.

For every integer $i \geq 0$, let S_i be the sphere centered at x_0 and of radius i. Note that S_i is a finite set of vertices of X. For $i \geq 1$, and for every subset V of S_i, we define $P(V)$ to be the set $\{x \in S_{i-1} \mid dist(x, V) = 1\}$, and we call $P(V)$ the *projection* of V on the sphere S_{i-1}. Let us note that if V is nonempty, $P(V)$ is also nonempty. Define $P^0(V) = V$, and by induction on $n = 1, 2, ...i$, $P^n(V) = P(P^{n-1}(V))$. $P^n(V)$ is a subset of the sphere S_{i-n} and is called the *projection* of V on that sphere.

Let us begin by a general lemma on the convexity of balls in δ-hyperbolic metric spaces:

Lemma 3.1. — *In a geodesic δ-hyperbolic space, balls are 4δ-convex.*

(We recall that this means that if $[x, y]$ is a geodesic segment in X whose endpoints x and y are in a ball B, then the entire segment is contained in the 4δ-neighborhood of that ball.)

PROOF. Let x_0 be the centre of B. Consider a geodesic triangle $[x_0, x, y]$. The two sides $[x_0, x]$ and $[x_0, y]$ are contained in B. This triangle being 4δ-narrow (Chapter 1, Proposition 1.3), every point on the side $[x, y]$ is contained in the 4δ-neighborhood of the union of the other two, and is therefore contained in the 4δ-neighborhood of the ball B. ∎

Lemma 3.2. — *Let $V \subset S_i$ be a subset which satisfies the following condition:*

$$diam P^n(V) \leq 20\delta \text{ for all } 0 \leq n \leq min(i, [10\delta + 1]).$$

Then,

$$diam P^n(V) \leq 20\delta \text{ for all } 0 \leq n \leq i.$$

PROOF. Let k be the integer $[10\delta + 1]$. It is sufficient to show that if $i \geq k + 1$, then the diameter of the set $P^{k+1}(V)$ is $\leq 20\delta$.

Let y_1 and y_2 be two points of $P^{k+1}(V)$. Then, there exist two points x_1 and x_2 in V such that

$$\mid y_1 - x_1 \mid = \mid y_2 - x_2 \mid = dist(x_1, S_{i-k-1}) = dist(x_2, S_{i-k-1}) = k + 1.$$

We have therefore $y_1 = p(x_1)$ and $y_2 = p(x_2)$, where p denotes here a projection operator on the quasi-convex subset $S_{i-(k+1)}$ of X (in the sense of [CDP], Chapter 10 §2; we recall that a map $p : X \to Y \subset X$ is called a "projection" if for every $x \in X$, we have $\mid x - p(x) \mid = dist(x, Y)$).

We apply w proposition 2.1 of Chapter 10 of [CDP], which expresses the contraction property of such a projection. We obtain:

$$\mid y_1 - y_2 \mid \leq max(20\delta, \mid x_1 - x_2 \mid - \mid x_1 - y_1 \mid - \mid x_2 - y_2 \mid + 20\delta) = 20\delta,$$

which shows that the diameter of $P^{k+1}(V)$ is also bounded above by 20δ. ∎

We define now a connected projective sequence of sets

$$... \to E_{i+1} \to E_i \to ... \to E_1 \to E_0$$

in the following manner:

For every $i = 0, 1, 2, ...$, the elements of E_i are nonempty subsets of the sphere $S_i \subset X$ which satisfy the hypothesis of Lemma 3.2.

The map $E_{i+1} \to E_i$ associates to such a nonempty subset V of S_{i+1} its projection $P(V)$ on the sphere S_i. Lemma 3.2 shows that this projection is an element of E_i.

We define then $T_{part} = T_{part}(X)$ to be the pointed tree associated to this projective sequence.

Let us now equip the tree T_{part} with its canonical metric.

Let ξ be a point on the boundary ∂T_{part} of T_{part}, and let γ be the unique geodesic ray in T_{part} starting at x_0 and converging to ξ. Let us call $\gamma(0) = V_0, \gamma(1) = V_1, \gamma(2) = V_2, \dots$ the consecutive vertices on γ. For every $i \geq 0$, the vertex V_i is represented by a finite subset V_i of the sphere $S_i \subset X$, and the sequence V_i has the property that for every $i \geq 0$, $V_i = P(V_{i+1})$.

We construct, from the sets $V_i \subset X$, a sequence of points $(x_i)_{i \geq 0}$ in X, by induction on i:

- x_0 is the basepoint,

- for every $i \geq 0$, $x_{i+1} \in V_{i+1}$ and $\mid x_i - x_{i+1} \mid = 1$.

Let us note that such a sequence x_i exists but that it is not necessarily unique.

For every $i \geq 0$, let us join the points x_i and x_{i+1} by a geodesic segment in X (of length 1) and let r be the total path which is thus defined. The origin of this path is x_0, and we parametrize it by arclength.

Proposition 3.4. — *The path $r : [0, \infty[\to X$ is a geodesic ray.*

PROOF. It suffices to see that for all $i \geq 0$, the subpath of r which is comprised between the origin $r(0)$ and the point $r(i) = x_i$ is geodesic. For this, we note that this subpath is of length i, and x_i is on the sphere S_i, which implies that $\mid x_i \mid = i$. This proves the proposition. ∎

The path r defines in this way a point $r(\infty)$ in ∂X. We have the following

Proposition 3.5. — *The point $r(\infty)$ is canonically associated to ξ (that is, this point does not depend on the choice of the sequence of points x_i).*

PROOF. As the diameters of the sets V_i are bounded above by 20δ, two geodesic rays r and r' which are constructed using two different choices of sequences (x_i) and (x'_i) of points in V_i are at a uniformly bounded distance from each other, and define therefore the same point at infinity. This proves the proposition. ∎

We now define the map $f_{part} : \partial T_{part} \to \partial X$, which to every point ξ of ∂T_{part} associates the point $r(\infty)$ of ∂X given by the construction above.

We equip the spaces ∂T_{part} and ∂X of their metrics $\mid \mid_a$ defined as in §2 above.

Proposition 3.6. — *The map $f_{part} : \partial T_{part} \to \partial X$ is Lipschitz (and therefore continuous) and surjective.*

PROOF. Let ξ and ξ' be two elements of ∂T_{part} and let γ and γ' be the geodesic rays in T_{part} which start at $\{x_0\}$ and which converge to these points. Let (x_i) and (x'_i)

be the sequences of points in X which are associated to the rays γ and γ' as above, and let r and r' be the two associated geodesic rays in X, with $r(\infty) = f_{part}(\xi)$ and $r'(\infty) = f_{part}(\xi')$. If the rays γ and γ' coincide along a distance t, we can take $x_i = x'_i$ for every $i \leq t$. Thus, the geodesic rays r and r' coincide along a distance equal to t. The fact that the map f_{part} is Lipschitz can be proved in the same manner as for the map f_{geo} in Proposition 2.3.

For the surjectivity, let ζ be an arbitrary element of ∂X. We shall construct a geodesic ray γ in the tree T_{part}, such that $\zeta = f_{part}(\gamma(\infty))$. For that, let R be the union of all the geodesic rays in X which originate at x_0 and which converge to ζ. We know that R is nonempty. For every $i \geq 0$, we define $V_i = R \cap S_i$. By one of the basic properties of δ-hyperbolic spaces (*cf.* inequality (3.4.1) of Chapter 1), we know that if r and r' are two geodesic rays in X which originate at x_0 and which converge to the same point ζ, we have, for all $i \geq 0$, $| r(i) - r'(i) | \leq 4\delta$. The diameter of every set V_i is therefore bounded above by 4δ. On the other hand, we have, for every $i \geq 1$, $P(V_i) = V_{i-1}$. Indeed, every point $x \in V_{i-1}$ is on a geodesic ray r contained in R. If $x' = r \cap S_i$, we have $| x - x' | = 1$ and $x' \in V_i$. Therefore $V_{i-1} \subset P(V_i)$. Conversely, let x be now a point of $P(V_i)$, and let x' be a point of V_i which satisfies $| x' - x | = 1$. The point x' belongs to a geodesic ray r originating at x_0 and satisfying $r(\infty) = \zeta$. Let g be a geodesic segment in X joining the basepoint x_0 to x, and let r' be the path defined by concatenation of g, of the geodesic segment $[x, x']$ and of the subpath r' of r comprised between x' and ζ. The path r', parametrized by arclength, is clearly geodesic, which shows that x belongs to the set V_{i-1}. Hence $P(V_i) = V_{i-1}$. The sequence (V_i) defines therefore a geodesic ray γ in the tree T_{part}, such that $f_{part}(\gamma(\infty))$ is the point $\zeta \in \partial X$. This shows that f_{part} is surjective. ∎

We suppose now that the graph X has bounded geometry, which means that there exists a constant N_0 which is an upper bound for the number of edges containing any vertex of X.

Proposition 3.7. — *Under the above hypothesis, the map* $f_{part} : \partial T_{part} \to \partial X$ *is finite-to-one. More precisely, we have,* $\forall \zeta \in \partial X$,

$$card(f_{part}^{-1}(\zeta)) \leq 2^{N_0^{24\delta+1}}.$$

PROOF. Let $g : [0, \infty[\to X$ be a geodesic ray with $g(0) = x_0$ and $g(\infty) = \zeta$, let $y_i = g(i)$, $i = 0, 1, 2, \ldots$ be the sequence of consecutive vertices on this ray, and for every i, let B_i be the closed ball in X centered at y_i and of radius 24δ.

Consider the sequence $V_0, V_1, V_2, \ldots$, where $V_0 = \{x_0\}$ and where for every $i \geq 0$, V_i is a subset of the sphere $S_i \subset X$ satisfying $V_i = P(V_{i+1})$. This sequence defines a geodesic ray in the tree T_{part}, and we suppose that this ray converges to a point $\xi \in \partial T_{part}$ satisfying $f_{part}(\xi) = \zeta$. We associate to the sequence (V_i), as above, a sequence of points $x_i \in V_i$ which is the sequence of consecutive vertices on a geodesic ray starting at x_0 and converging to ζ. We thus have $| x_i - y_i | \leq 4\delta$ for every $i = 0, 1, 2, \ldots$. As $diam V_i \leq 20\delta$, this implies that for every $i = 0, 1, 2, \ldots$, V_i is a subset of B_i.

Consider now another sequence $V_0', V_1', ... V_n', ...$ of subsets of X, with $V_0' = \{x_0\}$, and $V_i' = P(V_{i+1}') \; \forall i \geq 0$, defining, as the sequence in the tree T_{part}, a geodesic ray which originates at x_0.

If for a given i, we have $V_{i+1} = V_{i+1}'$, then we have also $V_i = V_i'$ (this is an elementary property of projections). We deduce that if the two sequences (V_i) and (V_i') are distinct, then V_i and V_i' are distinct subsets of X for every i large enough.

Let us suppose now that the two geodesic rays in T_{part} which we have just considered converge to the point ζ. We have seen that this implies that for all $i = 0, 1, 2, ...,$ $V_i \subset B_i$.

Let N_0 be, as above, an upper bound for the number of edges in X which contain an arbitrary vertex. Without loss of generality, suppose that $N_0 \geq 2$. Then, the number of distinct points in a ball B_i is bounded above by

$$N_1 = 1 + N_0 + N_0^2 + ... + N_0^{[24\delta]} \leq N_0^{24\delta+1}.$$

On the other hand, the number of distinct subsets of a set of cardinality N is equal to 2^N. Thus, the number of distinct subsets in a ball B_i is bounded above by $M = 2^{N_0^{24\delta+1}}$. The number of sequences (V_i) which are associated to the point $\zeta \in \partial X$ is therefore bounded above by M, which completes the proof of the proposition. ∎

Examples. In Figure 4, we have represented on the left hand side three graphs X. The trees $T_{part}(X)$ which are associated to these graphs are represented on the right hand side.

§4 – The tree T_{lex} associated to a hyperbolic group

In this section, Γ is a hyperbolic group together with a finite generating set G which does not contain the identity element Id, and X is the associated Cayley graph. We suppose that this graph is δ-hyperbolic. We recall that by definition of this graph, the set of vertices of X is the group Γ itself, and that two vertices x and $y \in X$ are joined by an edge if and only if, as elements of Γ, we can write $y = xg$ or $x = yg$, with $g \in G$. We take as a basepoint x_0 of X the identity element Id of the group.

Let $G^{-1} = \{g^{-1} \in \Gamma \mid g \in G\}$, and let $G' = G \cup G^{-1}$. If x is an arbitrary vertex of X, the set of edges which start at x is in a natural one-to-one correspondence with the set G' (a vertex which starts at x can be written in a unique way as xg, with $g \in G'$).

Let us fix an arbitrary total order relation on G'. For every vertex $x \in X$, this induces an order relation on the set of edges starting at x, and an order relation, called the "lexicographical ordering", on the set of all finite or infinite simplicial paths starting at the vertex x. We have the following proposition (whose proof is trivial):

Proposition 4.1. — *If $\mathcal{D}$ is an arbitrary set of paths starting at x, there is one and only one of these paths $\gamma \in \mathcal{D}$ which is smaller than all the other paths in $\mathcal{D}$ (in other words, the set of simplicial paths starting at x is well-ordered).* ∎

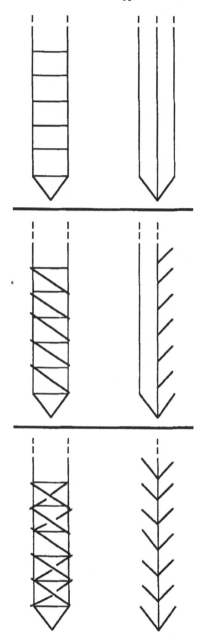

Figure 4

We now associate to X a pointed tree $T_{lex} = T_{lex}(X)$, which is the tree of geodesic simplicial paths which are smallest in the lexicographical order. We can define this tree in the same way in which we defined the tree $T_{geo}(X)$ in §2, by taking here as a set E_i the set of geodesic paths $r : [0, i] \to X$ which start at x_0 and which are smallest in the set of all geodesic paths joining the two endpoints $x_0 = r(0)$ and $r(i)$.

We note that $T_{lex}(X)$ is a pointed sub-tree of $T_{geo}(X)$. Let f_{lex} be the restriction of f_{geo} to $\partial T_{lex}(X) \subset \partial T_{geo}(X)$.

The map $f_{lex} : \partial T_{lex}(X) \to \partial X$ has a canonical section $s : \partial X \to \partial T_{lex}(X)$. Indeed, let ξ be an arbitrary point of ∂X. The set of geodesic rays starting at x_0 and terminating at ξ is well-ordered (for the lexicographical order on infinite rays), and therefore it possesses a smallest element. Let $r : [0, \infty[\to X$ be that element. The ray r has the property that for any integer $i \geq 0$, its restriction to $[0, i]$ is a finite geodesic segment, which is smallest in the set of geodesic segments which have the same endpoints. Thus, r defines an element of $\partial T_{lex}(X)$ whose image by f_{lex} is equal to ξ. The map s is then defined by $s(\xi) = r$.

The spaces ∂T_{lex} and ∂X are equipped with their metrics $\mid \mid_a$, for $a \in]1, a_0(\delta)[$.

Proposition 4.2. — *The map $f_{lex} : \partial T_{lex}(X) \to \partial X$ is Lipschitz (and therefore continuous), surjective and finite-to-one. More precisely, for every $\xi \in \partial X$, we have $card(f_{lex}^{-1}(\xi)) \leq card(B^0)$, where B^0 is the set of vertices in a closed ball $B = B_{4\delta}$ of radius 4δ centered at an arbitrary vertex of X.*

Let us remark that if $n = card(G \cup G^{-1})$, then we have $card(B^0) \leq n^{4\delta+1}$ (*cf.* Proposition 3.7).

PROOF. The map f_{lex} is Lipschitz because f_{geo} is Lipschitz. Furthermore, the existence of the section $s : \partial X \to \partial T_{lex}(X)$ shows that f_{lex} is surjective.

It remains to show that f_{lex} is finite-to-one.

Let $\xi \in \partial X$, and let $\mathcal{C}$ be the set of geodesic rays $r : [0, \infty[\to T_{lex}(X)$ whose endpoint is in the set $f^{-1}(\xi)$. If r is an arbitrary element of $\mathcal{C}$, it defines, by composition with the map $f'_L : T_{lex} \to X$, a geodesic ray $r' : [0, \infty[\to X$ which starts at x_0 and which converges to ξ. If r_1 and $r_2 : [0, \infty[\to T_{lex}(X)$ are two distinct elements of $\mathcal{C}$, with r'_1 and $r'_2 : [0, \infty[\to X$ their images by f'_L, the intersection of these two images in X is a connected set containing the origin x_0; see Figure 5. Indeed, in the contrary case, we would find two distinct geodesic segments which are smallest in the lexicographical order and which have their endpoints in common, which is not possible.

Suppose now that there exist N distinct geodesic rays in $\mathcal{C}$, which are denoted by $r_1, ..., r_N : [0, \infty[\to T_{lex}(X)$. Their images, $r'_1, ..., r'_N : [0, \infty[\to X$ converge to ξ, and according to the remark which follows Corollary 3.4 of Chapter 1, these image rays are all uniformly 4δ-close from each other.

Let us take an arbitrary vertex p on the image of the ray r'_1, and let us consider the closed ball $B = B_{4\delta}(p)$ of radius 4δ and center p. For all $i = 1, ..., N$, we have

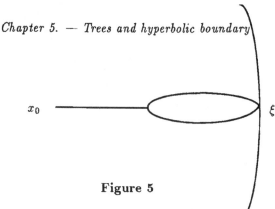

Figure 5

$Im(r_i') \cap B^0 \neq \emptyset$, and if p is close enough to ξ, the intersection points of the images of the r_i''s with B^0 are all distict. The number of such intersection points is therefore bounded by $card(B^0)$. This proves the proposition. ∎

Example. We can see, by the following example, that the map $f_{lex} : \partial T_{lex}(X) \to X$ is not necessarily injective.

Let us take $\Gamma = \mathbb{Z} \oplus \mathbb{Z}/2$ and $G = \{(0,1),(1,0)\}$. We then have $G \cup G^{-1} = \{(0,1),(1,0),(-1,0)\}$. We equip this last set with the following order relation: $(0,1) < (1,0) < (-1,0)$. The Cayley graph X is then represented by Figure 6 below, and we can see on that figure two geodesic rays which converge to the same point in ∂X but not to the same point in $\partial T_{lex}(X)$.

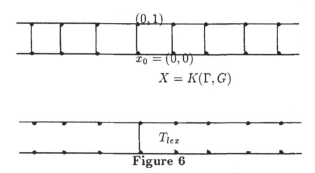

$$X = K(\Gamma, G)$$

Figure 6

§5 – A remark on the dimension of the hyperbolic boundary

The notion of *topological dimension* for a topological space has a long history (we refer the reader to the following books: [Fed], [HW], [Kur] and [Nag]). Let us recall (*cf.* Chapter 2) that a topological space has dimension 0 if and only if it has a basis for its topology whose elements are at the same time closed and open sets. From this definition, we can easily see that the boundary of a tree is a space which has topological dimension 0. For every integer $n \geq 1$, we define, by induction, the notion of topological dimension being $\leq n$. This is done in the following manner: a space is said to be of topological dimension $\leq n$ if its topology has a basis of open sets whose boundaries are of dimension $\leq n-1$ (*cf.* [HW]). Finally, we say that the space is *of dimension* n if it is of dimension $\leq n$ but not of dimension $\leq n - 1$. We have the following result, whose proof can be found in [Kur], Chapter XIX:

Theorem 5.1. — *Let n be a natural integer, X a compact metrizable space and T a compact topological space of dimension 0. Suppose that there exists a continuous map $f : T \to X$ which is surjective and which has the following property:*

$$card\big(f^{-1}(x)\big) \leq n + 1 \ \forall x \in X.$$

Then X has topological dimension $\leq n$. ∎

Thus, if X is a hyperbolic graph which has bounded geometry (for instance the Cayley graph of a hyperbolic group), Proposition 1.7 (and 4.2) gives us an upper bound for the topological dimension of ∂X. In particular, we have the following

Corollary 5.2. — *The boundary of a hyperbolic group is a topological space of finite dimension.*

Let us remark that the map π_0 defined in §8 of Chapter 3 gives us in the same manner an upper bound for the topological dimension of the boundary of a hyperbolic group.

Let us note also that in [Coo], there is an upper bound for the Hausdorff dimension of the boundary of a hyperbolic group (with respect to the metric $|\ |_a$). We know on the other hand that the topological dimension of a space is equal to the greatest lower bound of the set of Hausdorff dimensions of that space (for all the metrics which are compatible with its topology). This gives us another way of bounding the topological dimension of the boundary.

Finally, we note that M. Bestvina and G. Mess have computed the topological dimension of $\partial\Gamma$, for Γ hyperbolic, in terms of the cohomology of Γ (*cf.* [BM]).

Notes and comments on Chapter 5

This chapter develops certain ideas contained in sections 7.6, 8.5.B and 8.5.C of [Gro 2]. The trees which we have denoted by $T_{geo}(X)$, $T_{part}(X)$ and $T_{lex}(X)$ correspond to those which Gromov denotes respectively by T, $T(X)$ and $\overleftarrow{T}$. For the tree $T_{part}(X)$, we were led to modify the definition of Gromov so that this tree satisfies the required properties.

Bibliography for Chapter 5

[BM] M. Bestvina, G. Mess, "The boundary of negatively curved groups",Journal of the AMS, **4** (1991, pp. 469-481.

[Coo] M. Coornaert, "Mesures de Patterson-Sullivan sur le bord d'un espace hyperbolique au sens de Gromov" (to appear in the Pacific J. of Math.)

[CDP] M. Coornaert, T. Delzant, A. Papadopoulos, "Geométrie et théorie des groupes: les groupes hyperboliques de Gromov", Lecture Notes in Mathematics, vol. 1441, Springer Verlag, 1990.

[Fed] V. V. Fedorchuk, "The fundamentals of dimension theory", *in* "Encyclopaedia of mathematical sciences, General Topology, vol. I" Edit: A. V. Arckhangel'skii & L. S. Pontryagin, Springer Verlag (1990), pp. 91-194.

[Gro 1] M. Gromov, "Hyperbolic manifolds, groups and actions", *Ann. of Math. Studies* **97**, Princeton University Press, 1982 , pp. 183-215 .

[Gro 2] ——, "Hyperbolic groups", *in* Essays in Group Theory, MSRI publ. **8**, Springer Verlag, 1987, pp. 75-263.

[HW] W. Hurewicz & H. Wallman, "Dimension theory", Princeton University Press, 1948.

[Kur] K. Kuratowski, "Introduction à la théorie des ensembles et à la topologie", Monographie No. 15 de l'Enseignement Mathématique, Genève 1966.

[Nag] J. Nagata, "Modern dimension theory", Helderlann Verlag, Berlin, 1983.

[Ser] J-P. Serre, "Arbres, amalgames, SL_2", *Astérisque* **46**, 1977.

Chapter 6

Semi-Markovian spaces

In this chapter, we introduce *semi-Markovian* spaces. This is a special class of compact metrizable spaces. (In fact, we shall see that there are only countably many semi-Markovian spaces, up to homeomorphism). As a first example, we show that any finite simplicial complex is semi-Markovian. Other examples which we shall give include some "fractal" topological spaces such as the Cantor set, the Sierpinski sets and the Menger curve.

A semi-Markovian space admits, by definition , a *semi-Markovian presentation* which allows one to construct the space out of a finite set of combinatorial data. We shall see that there is a certain analogy between the notion of semi-Markovian presentation of a topological space and the notion of finite presentation of a dynamical system, which was introduced in Chapter 2. We note however that the notion of semi-Markovian presentation does not involve any group (or even semigroup) action.

The notions introduced in this chapter will be used in Chapter 7, where we will show that the boundary of a torsion-free hyperbolic group is semi-Markovian. This result will emphasize a certain "symbolic accessibility" of the boundary (as a topological space) of any torsion-free hyperbolic group.

§1 – Semi-Markovian spaces

Let S be a finite set and $\Sigma = \Sigma(\mathbb{N}, S)$ the one-sided Bernoulli shift with set of symbols S.

Definition 1.1. — We say that a subset Ψ of Σ is *semi-Markovian* if there exists a cylinder $C \subset \Sigma$ and a subshift of finite type $\Phi \subset \Sigma$ such that $\Psi = C \cap \Phi$.

The following properties of semi-Markovian subsets follow immediately from the analogous properties of cylinders and subshifts of finite type.

Proposition 1.2. — *The interesection of two semi-Markovian subsets of Σ is a semi-Markovian subset.* ∎

Proposition 1.3. — *If Ψ is a semi-Markovian subset of Σ, then the diagonal $\Delta(\Psi)$ of Ψ, defined as*

$$\Delta(\Psi) = \{(\psi, \psi) | \psi \in \Psi\}$$

is a semi-Markovian subset of $\Sigma \times \Sigma = \Sigma(\mathbb{N}, S \times S)$. ∎

Proposition 1.4. — *Let S_1 and S_2 be finite sets. Let Ψ_1 be a semi-Markovian subset of $\Sigma(\mathbb{N}, S_1)$ and Ψ_2 a semi-Markovian subset of $\Sigma(\mathbb{N}, S_2)$. Then $\Psi_1 \times \Psi_2$ is a semi-Markovian subset of the product Bernoulli shift*

$$\Sigma(\mathbb{N}, S_1) \times \Sigma(\mathbb{N}, S_2) = \Sigma(\mathbb{N}, S_1 \times S_2).$$

∎

Definition 1.5. — Let Ω be a Hausdorff topological space. We say that Ω is *semi-Markovian* if there exists a finite set S, a semi-Markovian subset $\Psi \subset \Sigma(\mathbb{N}, S)$ and a map $\pi : \Psi \to \Omega$ which is continuous and surjective, such that the equivalence relation associated to π, i.e.

$$R(\pi) = \{(\psi_1, \psi_2) \in \Psi \times \Psi | \pi(\psi_1) = \pi(\psi_2)\},$$

is a semi-Markovian subset of $\Sigma(\mathbb{N}, S) \times \Sigma(\mathbb{N}, S) = \Sigma(\mathbb{N}, S \times S)$. Such a map π is called a *semi-Markovian presentation* of Ω.

Remark. If $\pi : \Psi \to \Omega$ is as in the preceding definition, then, there is a quotient map $\Psi/R(\pi) \to \Omega$. By compactness of Ψ, the quotient $\Psi/R(\pi)$ is also compact, and the map $\Psi/R(\pi) \to \Omega$ is a continuous bijection from a compact set to a Hausdorff space; therefore , it is a homeomorphism. Thus, we have the following

Proposition 1.6. — *A semi-Markovian space is compact and metrizable.* ∎

For a given finite set S, it is clear that the set of cylinders (resp. subshifts of finite type) of $\Sigma(\mathbb{N}, S)$ is countable. The set of semi-Markovian subsets of $\Sigma(\mathbb{N}, S)$ is therefore also countable. Now using for instance the fact that any finite set, with the discrete topology, is semi-Markovian, we have the

Proposition 1.7. — *The class of semi-Markovian spaces, up to homeomorphism, is infinite countable.* ∎

On the other hand, we can easily produce an uncountable class of compact metrizable spaces such that any two elements of this class are nonhomeomorphic. Here is an example:

To every sequence $u = (u_n)_{n \geq 1}$ of natural integers, we associate the infinite product of spheres

$$\Omega_u = S^1 \times ... \times S^1 \times S^2 \times ... \times S^2 \times S^3 \times ...$$

where each sphere S^n, $n = 1, 2, ...$, appears u_n times. The space Ω_u is compact and mertizable, since it is the product of countably many compact metrizable spaces. On the other hand, if two such spaces Ω_u and $\Omega_{u'}$ are homeomorphic, then the sequences u and u' are equal. Indeed, it is easy to show by induction that $u_n = u'_n$ by (consider the rank of the π_n of each of these spaces). Thus, the spaces Ω_u form an uncountable family of compact metrizable spaces, and any two of these spaces are non homeomorphic. Note that the Ω_n are all connected and locally connected.

We can also exhibit an uncountable family of topologically distinct spaces which are all zero-dimensional, in the following way (see [M-Z]). First, recall Cantor's definition of the successive derived sets $X^{(\alpha)}$ (where α is an ordinal) of a topological space X. The first derived set $D(X) = X^{(1)}$ is the set of accumulation points of X. Then the $X^{(\alpha)}$ are defined inductively by:

$$X^{(0)} = X$$

$$X^{(\alpha+1)} = D(X^{(\alpha)})$$

$$X^{(\alpha)} = \cap_{\xi < \alpha} X^{(\xi)} \text{ if } \alpha \text{ is a limit ordinal.}$$

Now let ω be the first infinite ordinal, and define X_α, for every countable ordinal α, as the set of ordinals which are less than or equal to ω^α. Note that the X_α are countable sets. For the order topology, the X_α are compact, metrizable and zero-dimensional. It is not hard to see that the derived set of order α of X_α consists of a single point. Hence $(X_\alpha)^{(\beta)} = \emptyset$ for $\alpha < \beta$, and therefore the X_α are all topologically distinct. As the set of all countable ordinals is uncountable, we get the desired family. Note that each space X_α can be imbedded in the Cantor set (as can be done for all zero-dimensional compact metrizable spaces), and therefore in the real line. Finally, note that each X_α appears as the boundary of a locally finite simplicial tree T_α. To see this, it suffices to imbed X_α in the boundary of the homogeneous tree of order three (in which each vertex is adjacent to exactly three edges) and take T_α to be the convex hull of X_α.

Corollary 1.8. — *There exist uncountably many compact metrizable spaces which are not semi-Markovian and such that any two of them are non homeomorphic.* ∎

Proposition 1.9. — *The cartesian product (resp. the disjoint union) of two semi-Markovian spaces is a semi-Markovian space.*

PROOF. Let $\pi_i : \Psi_i \to \Omega_i$, $\Psi_i \subset \Sigma(\mathbb{N}, S_i)$, be a semi-Markovian presentation of Ω_i $(i = 1, 2)$. Proposition 1.4 shows that $\pi_1 \times \pi_2$ is a semi-Markovian presentation of the cartesian product $\Omega_1 \times \Omega_2$. Let Ω (resp. S, resp. Ψ) be now the disjoint union of Ω_1 and Ω_2 (resp. of S_1 and S_2, resp. of Ψ_1 and Ψ_2). We have a natural inclusion $\Psi \subset \Sigma(\mathbb{N}, S)$. The map $\pi : \Psi \to \Omega$, defined by $\pi_{|\Psi_i} = \pi_i$ $(i = 1, 2)$, is a semi-Markovian presentation of Ω, as the reader can easily check. ∎

§2 – First examples of semi-Markovian spaces

Proposition 2.1. — *Let S be an arbitrary finite set. Any semi-Markovian subset Ψ of $\Sigma(\mathbb{N}, S)$ is a semi-Markovian space.*

PROOF. Using Proposition 1.3, we see that the identity map is a semi-Markovian presentation of Ψ. ∎

We have seen, in Chapter 2, that $\Sigma(\mathbb{N}, S)$ is a Cantor set provided that $card(S) \geq 2$. We therefore have the following

Corollary 2.2. — *Any Cantor set is a semi-Markovian space.* ∎

We shall use now the binary expansion of real numbers to describe a semi-Markovian presentation of the segment $[0, 1]$. Let us begin by recalling the definition and a few elementary properties of the binary expansion.

We take $S = \{0, 1\}$ as a set of symbols. Let $\Sigma = \Sigma(\mathbb{N}, S)$ and let $\pi : \Sigma \to [0, 1]$ be the map defined by

$$\pi(\sigma) = \sum_{i \in \mathbb{N}} \sigma(i)/2^{i+1}.$$

We say that σ is a *binary expansion* of $\pi(\sigma)$. The map π is continuous and surjective, but it is not injective. In fact, given $x \in [0, 1]$, we have

$$card(\pi^{-1}(x)) = 2 \text{ for } x \in]0, 1[\cap \mathbb{Z}[1/2], \text{ (i.e. for } x = p/2^n, \text{ where } 0 < p < 2^n),$$

and $card(\pi^{-1}(x)) = 1$ if not.

If x is an element of $]0, 1[\cap \mathbb{Z}[1/2]$, then $x = \pi(\sigma) = \pi(\sigma')$ where σ and σ' have the following property:

(*)   There exists an integer $i_0 \geq 0$ such that

$\sigma(i) = \sigma'(i)$ for every $i \leq i_0 - 1$,

$\sigma(i_0) = 0$ and $\sigma'(i_0) = 1$,

$\sigma(i) = 1$ and $\sigma'(i) = 0$ for every $i \geq i_0 + 1$.

Conversely, if the elements σ and σ' of Σ verify $(*)$, then $\pi(\sigma) = \pi(\sigma') \in$ $]0, 1[\cap \mathbb{Z}\,[1/2]$.

Proposition 2.3. — *The map $\pi : \Sigma \to [0, 1]$ is a semi-Markovian presentation of the segment $[0, 1]$.*

PROOF. It suffices to show that the set

$$R(\pi) = \{(\sigma_1, \sigma_2) \in \Sigma \times \Sigma \,|\, \pi(\sigma_1) = \pi(\sigma_2)\}$$

is a semi-Markovian subset of $\Sigma \times \Sigma = \Sigma(\mathbb{N}, S \times S)$. We note that $R(\pi)$ is the set of ordered pairs which are either of the form (σ, σ) where σ is an arbitrary element of Σ, or of the form (σ, σ') or (σ', σ) with σ and σ' satisfying property $(*)$ above.

Let W be the subset of cardinality 24 of $(S \times S)^3$ consisting of words of length 3 on the alphabet $S \times S$ which belong to the following list:

$$((s, s), (t, t), (u, u)) \text{ where } s, t, u \in \{0, 1\},$$

$$((s, s), (t, t), (0, 1)) \text{ where } s, t \in \{0, 1\},$$

$$((s, s), (0, 1), (1, 0)) \text{ where } s \in \{0, 1\},$$

$$((0, 1), (1, 0), (1, 0)),$$

$$((1, 0), (1, 0), (1, 0)),$$

$$((s, s), (t, t), (1, 0)) \text{ where } s, t \in \{0, 1\},$$

$$((s, s), (1, 0), (0, 1)) \text{ where } s \in \{0, 1\},$$

$$((1, 0), (0, 1), (0, 1)),$$

$$((0, 1), (0, 1), (0, 1)).$$

Let Φ be the subshift of finite type $(\Sigma \times \Sigma)_W$. Let us recall that Φ is the set of ordered pairs (σ_1, σ_2) such that

$$(\sigma_1, \sigma_2)(i, i+1, i+2) \in W \text{ for every } i \in \mathbb{N}.$$

Let C be the cylinder of $\Sigma \times \Sigma$ defined as the set of $(\sigma_1, \sigma_2) \in \Sigma \times \Sigma$ such that the word (of length two on $S \times S$) $(\sigma_1, \sigma_2)(0, 1)$ is neither the word $((0, 1), (0, 1))$ nor the word $((1, 0), (1, 0))$.

We can easily verify that $R(\pi) = \Phi \cap C$, which shows that $R(\pi)$ is a semi-Markovian subset of $\Sigma \times \Sigma$. ∎

The preceding proposition, together with Proposition 1.4, gives the following

Corollary 2.4. — *For any integer $n \geq 0$, the cube $[0,1]^n$ is a semi-Markovian space. In fact, $\pi^n : \Sigma^n \to [0,1]^n$ is a semi-Markovian presentation of $[0,1]^n$.* ∎

We obtain the circle S^1 after identification of the endpoints of the segment $[0,1]$. Let us use again the notations of Proposition 2.3. Let $p : [0,1] \to S^1$ be the canonical projection and let π' be the composed map $\pi' = p \circ \pi : \Sigma \to S^1$. We remark that $R(\pi') = \Phi$. The map π' is therefore a semi-Markovian presentation of S^1. Hence the

Proposition 2.5. — *The circle S^1 is a semi-Markovian space. In fact, $\pi' : \Sigma \to S^1$ is a semi-Markovian presentation of S^1.*

Remark. One should note that the map $\pi' : \Sigma \to S^1$ which we have just used is a finite presentation of the dynamical system $(S^1, \mathbb{N})$ where the action of $\mathbb{N}$ on $S^1 = \mathbb{R}/\mathbb{Z}$ is generated by the map $x \mapsto 2x \ (mod\ 1)$. More generally, if Ω is a compact metrizable space equipped with a continuous map $T : \Omega \to \Omega$ which generates a finitely presented dynamical system $(\Omega, \mathbb{N})$, then the space Ω is semi-Markovian (this is because any subshift of finite type is a semi-Markovian subset of the ambient space).

Exercise 1. For any integer $n \geq 1$, we construct the *Sierpinski set S_n* by deleting from the unit cube $C_n = [0,1]^n$ countably many disjoint open cubes, in the following manner:

At step $i = 1$ of the construction, we subdivide C_n into 3^n small cubes, by subdividing each of the factors $[0,1]$ of C_n in three parts: $[0,1/3]$, $[1/3,2/3]$ and $[2/3,1]$. We then remove from C_n the interior of the small central cube, that is, $]1/3,2/3[^n$.

Then at step $i = 2$, we subdivide in the same way each of the $3^n - 1$ remaining small cubes and we remove from each of them the interior of the central cube, and so on for $i = 2, 3, \ldots$.

The result is, by definition, the Sierpinski set S_n.

We note that S_1 is the triadic Cantor set. The set S_2 is called the *Sierpinski curve* or the *Sierpinski carpet*, and is described in [Sie]. The set S_3 is called the *Sierpinski sponge*. Figure 1 represents an approximation of the Sierpinski curve. Prove the following facts:

1) S_n is connected and locally connected for every $n \geq 2$.
2) The sets S_n are two-by-two non homeomorphic. (Hint: The sphere of dimension $n - 1$ imbeds topologically in S_n as the boundary of one of the removed cubes.)
3) S_n is semi-Markovian. (Hint: S_n is the set of points x in C_n with the following property: there exists no integer k such that if we write the components of x in base 3, the k-th digit after the decimal point of each of these components is equal to 1.)

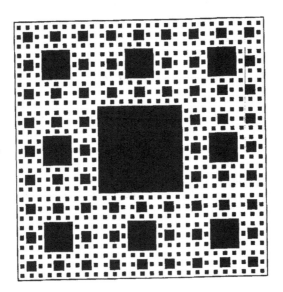

Figure 1

Exercise 2. The *Menger curve (or Menger sponge (cf.* [Men 1]) is the subset M of the unit cube of $\mathbf{R}^3$, defined as the set of points which project on each of the coordinate planes $x_i = 0$, into the Sierpinski curve. In other words, M is the set of points $x = (x_1, x_2, x_3) \in [0, 1]^3$ with the following property: for every $i, j = 1, 2, 3$, with $i \neq j$, and for every integer $k \geq 1$, the k-th digits after the decimal point of x_i and x_j, written in base 3, are not both equal to 1.

Use, as in the preceding exercise, the expression of the coordinates in base 3 to give a semi-Markovian presentation of M.

Remark: The Menger sponge and the Sierpinski sponge have very different topological properties. For example, the Sierpinski sponge is simply connected, whereas the Menger sponge is not.

§3 – Semi-Markovian presentation of polyhedra

By a *finite polyhedron*, we mean here a topological space which is homeomorphic to (the topological realization of) a finite simplicial complex. In this section, we establish the following theorem (whose proof follows almost immediately from Proposition 2.3).

Theorem 4.1. — *A finite polyhedron is a semi-Markovian space. In fact, if K is a finite polyhedron, then there exists a finite-to-one semi-Markovian presentation, that is, a presentation $\pi : \Psi \to K$ which satisfies the following property:*

(F) *There exists an integer N such that card $\left(\pi^{-1}(x)\right) \leq N$ for every $x \in K$.*

PROOF. We can easily transform a triangulation of K into a polyhedral decomposition in cubical cells (*i.e.* cells isomorphic to $[0,1]^n$, where $n = 0, 1, \dots$ depends on the cell considered). This can be done in the following manner. We begin by remarking with Gromov (*cf.* [Gro 3] pp. 131-132) that the cone of the first barycentric subdivision of an n-simplex can be seen, in a natural manner, as an $(n+1)$-cubical polyhedron. This allows us, using the first barycentric subdivision, to decompose any n-simplex into $n+1$ cubes of dimension n (Figure 2). Applying this construction to each one of the simplices of the triangulation, we obtain our cubical decomposition of K. (One can object that, in this construction, some of the cubes can meet along more than one face, but this has no importance for what follows.)

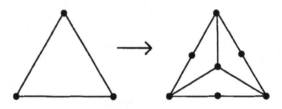

Figure 2

Let us show now how such a cubical decomposition allows us to construct a semi-Markovian presentation of K.

Let $C_1, \dots, C_p$ be maximal cubes of the decomposition (a cube is said to be *maximal* if it is not contained in any other one) . Each C_i can be identified with $[0,1]^n$, where $n = n(i)$ is the dimension of C_i. Let then $\pi_i : \Sigma_i \to C_i$ be the semi-Markovian presentation given in Corollary 2.4, where Σ_i is the one-sided Bernoulli shift with set of symbols $S_i = \{0,1\}^{n(i)}$, $i = 1, \dots, p$. Let S (resp. Ψ) be the disjoint union of the S_i (resp. of the Σ_i). We have $\Psi \subset \Sigma = \Sigma(\mathbb{N}, S)$. Consider the unique map $\pi : \Psi \to K$ that restricts to π_i on each Σ_i, for all i. Let us show that π is a semi-Markovian presentation of K.

114

We can easily see that Ψ is a subshift of order two of Σ. Indeed, Ψ is the set of maps $\sigma : \mathbf{N} \to S$ with the following property:

$$\exists i \in \{1,...,p\} \text{ such that } \forall j \in \mathbf{N}, \big(\sigma(j), \sigma(j+1)\big) \in S_i \times S_i.$$

In the same way, we show that

$$R(\pi) = \{(\psi_1, \psi_2) \in \Psi \times \psi \mid \pi(\psi_1) = \pi(\psi_2)\}$$

is a semi-Markovian subset of $\Sigma \times \Sigma$ (remark that the *standard* semi-Markovian presentation of a cube of dimension n, that is, the presentation given by Proposition 2.4, induces a standard presentation of each of the faces of the considered cube).

It remains to show that π satisfies property (F). This results from the fact that $card\big(\pi_i(x)\big) \leq 2^{dim\ C_i}$ for all $i = 1,...,p$ and for all $x \in C_i$. ∎

Corollary 4.2. — *A differentiable compact manifold is a semi-Markovian space.*

Indeed, according to a famous result of Cairns and Whitehead (see [Mun]), every compact differentiable manifold admits a triangulation (*i.e.* is homeomorphic to a simplicial complex).

Notes and comments on Chapter 6

Gromov introduces the notion of semi-Markovian space in section 8.5.H of [Gro 3].

All the connected semi-Markovian spaces given in this chapter are also locally connected, and it would be tempting to conjecture that any space Ω which is semi-Markovian and connected is also locally connected (see this conjecture, in the particular case where Ω is the boundary of a hyperbolic group in[BM], [Ghy] and [Mar]).

Note that the proof given for Corollary 1.8 does not produce any explicit example of a compact metrizable non semi-Markovian space. The following recent results of D. Fried allow us to construct very easily such examples. Fried [Fri2] proved that any semi-Markovian space has finite topological dimension. This implies with Theorem 4.1 that the space Ω_u described in §1 is semi-Markovian if and only if the sequence u has finite support. Fried also proved that for Ω semi-Markovian, there is an integer n such that $\Omega^{(n)} = \Omega^{(n+1)}$. This implies that the space X_α of §1 is semi-Markovian if and only if α is finite (see [Fri 3]).

The reader can find in [Ben] a description of hyperbolic groups whose boundary is homeomorphic to the Sierpinski curve (resp. to the Menger curve). In a certain sense, the Sierpinski curve (resp. the Menger curve) is the most complicated of all planar curves (resp. of all curves). Indeed, any planar curve (resp. any curve) imbeds topologically in the Sierpinski curve (resp. in the Menger curve). By "curve", we mean here a metrizable compact connected set of dimension one. For the proofs of these "universal" properties of the Sierpinski and the Menger curves, the reader is referred to [Bl Me], [Men 2] and [Sie].

Bibliography for Chapter 6

[Ben] N. Benakli, "Groupes hyperboliques de bord la courbe de Menger ou la courbe de Sierpinski", Thesis, Université de Paris-Sud (Orsay), 1992.

[BM] M. Bestvina et G. Mess, "The boundary of negatively curved groups", Journal of the AMS,4 (1991) pp.469-481.

[Bl Me] L. M. Blumenthal and K. Menger, "Studies in geometry", W. H. Freeman, 1970.

[Fri 2] D. Fried, "Dimension bounds for quotient spaces", preprint IHES, 1992.

[Fri 3] D. Fried, paper in preparation.

[Gro 3] M. Gromov, "Hyperbolic groups", *in* Essays in Group Theory, MSRI publ. **8**, Springer Verlag, 1987, pp. 75-263.

[Mar] I. Martinez, "Bord d'un produit amalgamé sur $\mathbb{Z}$ de deux groupes libres ou de surface", prépublication, 1991.

[Men 1] K. Menger, "Die Universalkurve", *Proc. Ac. Amsterdam*, **29** (1926) p. 1125.

[Men 2] K. Menger, "Kurventheorie", 2nd edition, Chelsea, 1967.

[MS] S. Masurkiewicz, W. Sierpinski, "Contribution à la topologie des ensembles dénombrables", *Fund. Math.*, **1** (1920) pp. 17-27.

[Mu] J. R. Munkres, "Elementary differential topology", Princeton University Press, 1963 .

[Sie] W. Sierpinski, "Sur une courbe cantorienne qui contient une image biunivoque et continue de toute courbe donnée", *C. R. A. S.* **162**, 1916 , pp. 629-632.

Chapter 7

The boundary of a torsion-free hyperbolic group as a semi-Markovian space

In this chapter, Γ is a torsion-free hyperbolic group. We show that the boundary, $\partial\Gamma$, is a semi-Markovian space, in the sense of Chapter 6. We construct semi-Markovian presentations of $\partial\Gamma$ using the trees T_{geo} and T_{part} which are associated to the Cayley graph of Γ with respect to a finite set of generators of Γ. These trees have been studied in Chapter 5. In the meantime, we shall describe the boundaries ∂T_{geo} and ∂T_{part} of T_{geo} and T_{part} as semi-Markovian subsets of one-sided Bernoulli shifts.

The set of symbols, for each of these presentations, is a set of "N-types". The N-type of a point of the group is a certain geometrical property of the ball of radius N centered at this point. There is always a finite number of N-types in a group (this is true for an arbitrary finitely generated group), and one fundamental property of hyperbolic groups is that the N-type at a point $x \in \Gamma$ determines in a certain sense the geometry of the set of points situated "beyond x" with respect to the origin. We shall express this property in a precise manner in terms of isometry classes of "pointed subtrees" of T_{geo} or of T_{part}.

The first part of this chapter is consacrated to these results on the N-type. We then express the boundary ∂T_{geo} of T_{geo} as a semi-Markovian subset of a certain one-sided Bernoulli shift. The map $f_{geo} : \partial T_{geo} \to \partial\Gamma$, which was constructed in Chapter 5, gives a semi-Markovian presentation of $\partial\Gamma$. We then show how to proceed in the same manner with the tree T_{part} and obtain a finite-to-one semi-Markovian presentation of $\partial\Gamma$.

§1 – N-equivalence in the trees T_{geo} and T_{part}

One important notion for a finitely generated group equipped with a word metric, is the notion of N-type at a point. This is a property of the ball of radius N centered at this point which depends not only on the geometry of the ball in itself, but also on the position of that ball with respect to the identity element of the group.

The notion of N-type originates in a paper of J. Cannon [Can]. The main properties described in [Can], concerning the N-type in fundamental groups of closed Riemannian manifolds of negative curvature can be easily generalized to hyperbolic groups (see for instance [CDP], Chapter 12). We recall below all the definitions and the properties which we shall need, but let us note right away that for any $N \geq 0$, there are finitely many N-types associated to differents points of a finitely generated group equipped with a word metric. Let us note also that one fundamental property of hyperbolic groups is that if N is large enough, the N-type determines the "cone type" (using the terminology of [Can], which is used also in [CDP]). The cone type at a point x of X is a property of the isometry class of the space of points of X which are "beyond" x with respect to the origin.

In all this chapter, Γ is a hyperbolic group equipped with a fixed finite set $G \subset \Gamma$ of generators and with the associated word metric and $\delta \geq 0$ is a constant such that the Cayley graph $K(\Gamma, G)$ is a δ-hyperbolic metric space.

For every integer $N \geq 0$ and for every element x of Γ, we denote by $B_N(x)$ the closed ball in Γ of centre x and radius N, and τ_x the unique left-translation in the group which sends the identity element Id to x (in other terms, τ_x is left-multiplication by x).

Definition 1.1. — (*N-type and N-equivalence in a group*). For every $N \in \mathbf{N}$ and $x \in \Gamma$, let $f_{N,x} : B_N(x) \to \mathbf{N}$ be the function which associates to every clement $z \in B_N(x)$ its distance to the identity element, $| z |$.

To this function, we can associate, in a canonical manner, a function $\bar{f}_{N,x}$ defined on the ball $B_N(Id)$ which takes the value 0 at the identity, by the formula

$$\bar{f}_{N,x} = f_{N,x} \circ \tau_x - |x|.$$

In other terms, we have

$$\forall z \in B_N(Id), \ \bar{f}_{N,x}(z) = |xz| - |x|.$$

We call $\bar{f}_{N,x}$ the N-type of x.

We shall say that two elements x and y of Γ are *N-equivalent* if they have the same N-type.

Note that an equivalent definition would be to say that the two elements x and y are N-equivalent if the left-translation γ of the group which sends x to y (*i.e.* left multiplication by $\gamma = yx^{-1}$) preserves up to a constant the function $f_{N,x}$, or, in other words, if the following condition is satisfied:

$$\exists C \in \mathbf{R} \mid \forall z \in B_N(x), |z| = C + |\gamma z|.$$

Taking $z = x$, we see that such a constant C, if it exists, is given by $C = |x| - |y|$.

Definition 1.2. — (*N-equivalent subsets*). This is a generalization of the preceding definition. Let B_1 and B_2 be two subsets of the group Γ. We say that they are *N-equivalent* if there exists a left-translation γ of the group such that $\gamma(B_1) = B_2$, and such that γ preserves the N-type at each point $x \in B_1$. In other words,

$$\forall x \in B_1, \exists C_x \in \mathbf{R} \text{ such that } \forall z \in B_N(x), |z| = C_x + |\gamma z|,$$

where for each x, C_x is a constant which depends only on x.

Note in particular that two sets which are N-equivalent have the same cardinality.

We denote respectively by T_{geo} and T_{part} the trees $T_{geo}(X)$ and $T_{part}(X)$ which were defined in Chapter 5, where we take as a space X the Cayley graph $K = K(\Gamma, G)$.

Definition 1.3. — (*N-equivalence in the trees T_{geo} and T_{part}*). Let x and y be two vertices of T_{geo}. To these vertices, there corresponds two well-defined elements of Γ (the endpoints of the corresponding geodesic segments). In the same way, if x and y are two vertices of the tree T_{part}, then there are two subsets of Γ which are naturally associated to them. We shall say that these two vertices of the tree T_{geo} (resp. T_{part}) are *N-equivalent* if the corresponding points (resp. subsets) in Γ are N-equivalent in the sense of Definition 1.1 (resp. 1.2).

Proposition 1.4. — *For every $N \geq 0$, the points of Γ are partitioned into a finite number of N-equivalence classes.*

PROOF. We have, for every $z \in B_N(Id)$,

$$|\bar{f}_{N,x}(z)| = \big| \, | \, xz \, | - | \, x \, | \, \big| \leq |xz - x| = |z| \leq N.$$

The set $B_N(Id)$ being finite, the number of N-types is also finite. ∎

Corollary 1.5. — *For every $N \geq 0$, the vertices of the tree T_{geo} are partitioned into a finite number of N-equivalence classes.* ∎

Proposition 1.6. — *For every $N \geq 0$, the vertices of the tree T_{part} are partitioned into a finite number of N-equivalence classes.*

PROOF. Up to the action of a group on itself by left multiplication, there are only finitely many subsets of that group which are of diameter $\leq 20\delta$. We can then use the same reasoning that we used in the proof of Proposition 1.4 to conclude the proof. ∎

The notion of a "pointed subtree" which we introduce now is intimately related to the notion of "cone type" in a group, introduced by Cannon in [Can], and which we have used in Chapter 12 of [CDP].

Let T be an arbitrary simplicial tree, equipped with a basepoint x_0. For each $x \in T$, we define the subtree $T_x \subset T$ as

$$T_x = \{y \in T \mid x \in [x_0, y]\}.$$

T_x is equipped with a natural basepoint (the point x itself). We say that T_x is the set of points T which are situated *beyond* x.

Definition 1.7. — (*N-equivalence of pointed subtrees*) Let x and y be two vertices of the tree $T = T_{geo}$ (resp. T_{part}). We say that the two pointed subtrees T_x and T_y are N-equivalent if the two points x and y are N-equivalent and if the left-translation γ which sends x to y induces an isometry between T_x and T_y.

The following two propositions imply that for N large enough, the second property in the above definition is a consequence of the first one:

Proposition 1.8. — *There exists an integer $N_0 \geq 0$ (which depends only on δ) such that, for every $N \geq N_0$, if x and y are two vertices of T_{geo} which are N-equivalent, then the pointed subtrees $T_{geo,x}$ and $T_{geo,y}$ are N-equivalent.*

PROOF. This is a fundamental result due to Cannon (*cf.* [Can] for the case where Γ is the fundamental group of a compact Riemannian manifold of negative sectional curvature, and [CDP] Chapter 12, for an adaptation to the case of an arbitrary hyperbolic group). For a proof, we refer the reader to this last reference . ∎

We have an analogous proposition for the vertices of T_{part}:

Proposition 1.9. — *If $N \geq N_0$ (the same N_0 as in the preceding proposition) and if x and y are two vertices of the tree T_{part} which are N-equivalent, then the two pointed subtrees $T_{part,x}$ and $T_{part,y}$ are N-equivalent.*

PROOF. We can adapt the proof of the result in Proposition 1.8 to the vertices of T_{part}, which are finite subsets of Γ. We can also deduce the result for the vertices of T_{part} from the corresponding result for the vertices of T_{geo} in the following manner:

Consider two vertices x and y of T_{part} which are N-equivalent, and denote in the same way by x and y respectively the two subsets of Γ which are represented by these two vertices. Let γ be the left-translation of the group which sends x to y, and which preserves the N-type at each point of x. Let us describe the pointed tree $T_{part,x} \subset T_{part}$ as a projective sequence of sets (as in §1 of Chapter 4):

$$... \to E_{i+1} \to E_i \to ... \to E_0,$$

and, in the same manner, let us define the pointed subtree $T_{part,y} \subset T_{part}$ as a projective sequence of sets

$$... \to E'_{i+1} \to E'_i \to ... \to E'_0.$$

For every element $z \in x$, we can define the projective sequence of sets which describes the pointed subtree $T_{geo,z} \subset T_{geo}$:

$$... \to E_{i+1}(z) \to E_i(z) \to ... \to E_0(z),$$

and for $z' \in y$, the one which describes the pointed subtree $T_{geo,z'} \subset T_{geo}$:

$$... \rightarrow E'_{i+1}(z') \rightarrow E'_i(z') \rightarrow ... \rightarrow E'_0(z').$$

For every $i \geq 0$, the set E_i (resp. E'_i), considered as a set of subsets of Γ, is the union of the sets $E_i(z)$ for z varying in x (resp. of the sets $E'_i(z')$ for z' varying in y). For every $z \in x$, the translation γ induces an isometry $\gamma^*(z)$ between the pointed trees $T_{geo,z}$ and $T_{geo,z'}$ (where $z' = \gamma z$) of T_{geo}. This tanslation induces in a natural way, for every $i \geq 0$, a bijection $f'_{i,z} : E_i(z) \rightarrow E'_i(z')$. Taking the unions, γ induces, for every $i \geq 0$, a bijection $f^*_i : E_i \rightarrow E'_i$. This last bijection induces in a natural way an isometry $f^* : T_{part,x} \rightarrow T_{part,y}$. In the same manner as for the maps $\gamma^*(z)$, and by the naturality of the construction, the isometry f^* is induced by the translation γ of the group. ∎

Corollary 1.10. — *For every $N \geq N_0$ (the same N_0 as in Proposition 1.8) and for each of the two trees T_{geo} and T_{part}, there are only finitely many N-equivalence classes of pointed subtrees (respectively of the form $T_{geo,x}$ and $T_{part,x}$).*

PROOF. This is a consequence of Propositions 1.8 et 1.9. ∎

§2 – The boundary of $\mathbf{T_{geo}}$ as a semi-Markovian subset

We suppose from now on that the group Γ is torsion-free. Recall then that all the elements of Γ are (with the exception of the identity element) of hyperbolic type. Recall also that Γ is equipped with a finite generating set $G \subset \Gamma$, and that $K = K(\Gamma, G)$ designates the Cayley graph associated to (Γ, G). An *oriented edge* of K is an ordered pair of adjacent vertices in this graph. It is a consequence of the definition of the Cayley graph of a finitely generated group that there is a natural labelling of the oriented edges of K, that is, a map from this set to the set $G \cup G^{-1}$: we label with the letter a the oriented edge $[x, xa]$ for all $x \in \Gamma$ and $a \in G \cup G^{-1}$.

The following proposition can be considered as the definition as the hyperbolicity for an element in the group (*cf.* Chapter 1, §7). We state it as such for later use.

Proposition 2.1. — *Let $\gamma \neq Id$ be an arbitrary element of Γ, γ being written as a finite product $g_1.....g_n$ of elements in $G \cup G^{-1}$. Let s be a simplicial path in the Cayley graph K, which is parametrized by arclength, and whose sequence of edges are labelled with the sequence $g_1.....g_n$ repeated l times,. Then, there exist constants λ, k and L (depending on γ, but not on l) such that the path s is a (λ, k)-quasi-geodesic, and such that if g is a geodesic path in K joining the two endpoints of s, each of the paths s and g is contained in the L-neighborhood of the other one.* ∎

Proposition 2.2. — *Let x be an arbitrary vertex of the tree T_{geo}, and let y_1 and y_2 be two distinct vertices of this tree which are situated beyond x (in the sense defined in §1 above, after Proposition 1.6) and at distance 1 from x. Then there exists an* integer N_1 *(which, for a given group* Γ, *depends only on the set G) such that for every* $N \geq N_1$, *the vertices y_1 and y_2 are not N-equivalent.*

PROOF. Let us denote in the same way the vertices x, y_1 and y_2 and their respective images in the Cayley graph, and let a and b be respectively the elements of $G \cup G^{-1}$ which correspond to the two oriented edges $[y_1, x]$ and $[x, y_2]$ (see Figure 1).

Figure 1

$| x |$ being the distance from x to the identity element Id, we have:

$$| y_1 | = | y_2 | = | x | + 1.$$

Consider now, in the Cayley graph, the simplicial path s of length $2N$, with origin the point y_1 and which as a word in the alphabet $G \cup G^{-1}$, is written as $ababab...ab$ (ab repeated N times) (see Figure 2).

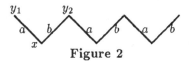

Figure 2

Let us suppose that the verices y_1 and y_2 are $2N$-equivalent. Applying the definition of the $2N$-equivalence, we see that the consecutive vertices in the path s are alternatively at distance $| x |$ and $| x | + 1$ from the identity element Id of the group.

The assumption on the group Γ that it is torsion-free implies that the word ab, regarded as an element of Γ, is of hyperbolic type. Thus, there exist constants λ and k such that the path s is a (λ, k)-quasi-geodesic (where λ and k depend on ab).

Let y_3 be the terminal point of the path s, $[y_1, y_3]$ a geodesic segment joining the points y_1 and y_3, and $[Id, y_1]$ and $[Id, y_3]$ two geodesic segments joining the identity element to the points y_1 and y_3. Let Δ be the geodesic triangle defined as the union of the three segments $[y_1, y_3]$, $[Id, y_1]$ and $[Id, y_3]$, and z the midpoint of the segment $[y_1, y_3]$. The point z is at distance $\leq L$ from a point on the path s (where L is the constant given by Proposition 2.1), and we therefore have

(2.2.1) $| z - Id | \geq | x | - L.$

The length of the path s is equal to $2N$. As this path is (λ, k)-quasi-geodesic, we have

$$| y_1 - y_3 | \geq 2N/\lambda - k.$$

123

By δ-hyperbolicity, the triangle Δ is 4δ-thin, and the point z is therefore at distance $\leq 4\delta$ from a point z' on the union $[Id, y_1] \cup [Id, y_3]$. Let us suppose that z' is on the segment $[Id, y_i]$, where $i = 1$ or 3. We have then

$$| z' - y_i | \geq | z - y_i | - | z - z' |$$

$$\geq 1/2(2N/\lambda - k) - 4\delta = N/\lambda - k/2 - 4\delta.$$

We deduce that

$$| z' - Id | \leq | y_i | - N/\lambda + k/2 + 4\delta.$$

Now, using (2.2.1) and the fact that $| y_i | = | x | + 1$, we have:

$$| x | - L - 4\delta \leq | x | + 1 - N/\lambda + k/2 + 4\delta,$$

which gives

$$N \leq \lambda(1 + L + k/2 + 8\delta).$$

Thus, the integer N is bounded above by a constant which depends only on the word ab. On the other hand, there are only finitely many elements γ in the group which are of the form $\gamma = ab$, with a and b varying in $G \cup G^{-1}$. Taking N_1 larger than all the constants $2N$ associated to the different possible choices of ab, we see that two arbitrary vertices y_1 and y_2 which satisfy the hypothesis of the proposition cannot be N'-equivalent, for any $N' \geq N_1$. This proves the proposition. ∎

Consider now the tree T_{geo} associated to (Γ, G). Given an integer $N \geq 0$, let S be the set of N-equivalence classes of vertices of T_{geo} (*cf.* Definition 1.3). The set S is finite (Corollary 1.5). In what follows, we take N to be at least equal to the integer N_0 of Proposition 1.8 and to the integer N_1 of Proposition 2.2.

The set S will be the set of symbols of the semi-Markovian presentation which we are looking for. Let $\Sigma = \Sigma(\mathbb{N}, S)$ be the one-sided Bernoulli shift associated to S.

Definition 2.3. — Let s_1 and s_2 be two elements of S. We say that s_2 *follows* s_1 *in the tree* T_{geo} if s_1 and s_2 can be represented respectively by vertices x' and y' of this tree, where the point y' follows the point x' (in other terms, we have $y' \in T_{geo,x'}$ and $| x' - y' | = 1$).

The next proposition shows that the fact that s_2 follows s_1 is a property which can be verified at an arbitrary point x in T_{geo}, which is in the class s_1.

Proposition 2.4. — *Let* s_1 *and* s_2 *be two elements of* S *such that* s_2 *follows* s_1 *in* T_{geo}. *If* $x \in s_1$ *is an arbitrary vertex of* T_{geo} *in the class* s_1, *there exists a unique vertex* y *which follows* x *and which is in the class* s_2. *In fact, if* x' *and* y' *are as in the above definition, and if we identify (as in Definition 1.3) the four points* x, y, x' *and* y' *with their images in* Γ, *then* y *is the image of* y' *by the left-translation of the group* Γ *which sends* x' *on* x.

PROOF. By definition, we can find two vertices x' and y' of T_{geo}, where the point y' follows the point x', and such that x' and y' belong respectively to the classes s_1 and s_2. By Proposition 1.8, the trees $T_{geo,x}$ and $T_{geo,x'}$ are N-equivalent. As in the statement, let us denote by the same letters the three points of Γ which represent the three vertices x, y and y' of T_{geo}. Let γ be the left-translation of the group which sends x' on x. This translation preserves the N-type at x', and induces therefore (by Corollary 1.10) an isometry between the trees $T_{geo,x'}$ and $T_{geo,x}$. Let us call y the vertex $\gamma y' \in \Gamma$. As y' belongs to $T_{geo,x'}$ and as this point is situated at a distance 1 from this vertex, the same holds for the vertex y with respect to the vertex x , because γ send isometrically $T_{part,x'}$ on $T_{part,x}$. Thus, as elements of the tree T_{geo}, y follows x. Furthermore, γ preserves the N-type of y'. Indeed, we can classify the points of $B_N(y')$ in two categories, those which are in $B_N(x')$ and those which are not. The points of this last category can be considered as belonging to the tree $T_{geo,x'}$. We know that there exists a constant C such that for all $z \in B_N(x')$, we have $\mid z \mid = C + \mid \gamma z \mid$. This inequality is therefore verified for all the points of $B_N(y')$ which belong to the first category. As γ induces an isometry between the trees $T_{geo,x'}$ and $T_{geo,x}$, the points in the second category satisfy also the same inequality. Thus, y and y' have the same N-type. This proves Proposition 2.4. ∎

Let now Ψ_0 be the set of sequences of points in T_{geo} of the form $\big(g(n)\big)_{n\in\mathbb{N}}$, which are associated to geodesic rays in the tree and which can be written as $g(t)_{t\in[0,\infty[}$, with $g(0) = x_0$ (the basepoint of T_{geo}). In other terms, Ψ_0 is the set of "discrete geodesic rays" starting at the basepoint of the tree. Ψ_0 is identified to ∂T_{geo}, and we can define a map

$$P : \Psi_0 \rightarrow \Sigma = \Sigma(\mathbb{N}, S)$$

in the following manner:

If $g = \big(g(n)\big)_{n\in\mathbb{N}}$ is an element of Ψ_0, $P(g)$ is the element of Σ which associates to an integer $n \in \mathbb{N}$ the element of S that represents the N-equivalence class of the point $g(n)$.

Proposition 2.5. — *The map $P : \Psi_0 \rightarrow \Sigma$ is injective.*

PROOF. Let us take two elements $g = \big(g(n)\big)_{n\in\mathbb{N}}$ and $g' = \big(g'(n)\big)_{n\in\mathbb{N}}$ of Ψ_0 which have the same image by P. As $g(0) = g'(0) = x_0$, we have, by Proposition 2.2, $g(n) = g'(n)$ for every $n \in \mathbb{N}$. ∎

Let now $\Phi \subset \Sigma(\mathbb{N}, S)$ be the set of sequences $(s_i)_{i\geq 0}$ which satisfy the following condition:

$\forall n \geq 0$, the element s_{n+1} of S follows the element s_n in the sense of Definition 2.3.

Proposition 2.6. — *Φ is a subshift of finite type of $\Sigma(\mathbb{N}, S)$.*

PROOF. The definition of Φ shows that Φ is a subshift of order two of $\Sigma(\mathbb{N}, S)$. ∎

Let $\Psi \subset \Sigma(\mathbf{N}, S)$ be the image of Ψ_0 by the map P.

Proposition 2.7. — Ψ *is a semi-Markovian subset of* $\Sigma(\mathbf{N}, S)$.

PROOF. Ψ is the intersection of Φ with the cylinder

$$C = \{(s_n)_{n \in \mathbf{N}} \mid s_0 \text{ is the class of } x_0\}$$

(where x_0 designates as before the basepoint of the tree T_{geo}). ∎

Let now $P_0^{-1} : \Psi \to \Psi_0$ be the inverse map of P. The set Ψ_0 is in natural one-to-one correspondence with the boundary ∂T_{geo} of the tree T_{geo}. Let $D : \Psi_0 \to \partial T_{geo}$ be the natural map between these two spaces, and π_0 the map defined as the composition

$$\pi_0 = D \circ P_0^{-1} : \Psi \to \partial T_{geo}.$$

We have proved the following

Theorem 2.8. — *The map* $\pi_0 : \Psi \to \partial T_{geo}$ *is a homeomorphism between the semi-Markovian subset* $\Psi \subset \Sigma(\mathbf{N}, S)$ *and* ∂T_{geo}. *In particular, this map is a semi-Markovian presentation of* ∂T_{geo}. ∎

§3 – A semi-Markovian presentation of ∂Γ

In the rest of this chapter, we suppose for convenience (and without loss of generality) that the constant δ is an integer. For the semi-Markovian presentation of $\partial\Gamma$, we shall take as a set of symbols the same set S as in §2, that is, the set of N-equivalence classes of vertices of T_{geo}, for a convenient integer N which will be chosen later on. We begin by supposing that N satisfies the properties that we imposed in §2 (in other words, that it is at least equal to the integer N_0 of Proposition 1.8 and to the integer N_1 of Proposition 2.2), and we suppose also that $N \geq 4\delta$ and $\geq N_2$, where N_2 is the integer that will appear in Poposition 3.1 below.

Using the notations introduced in §2, we define the map

$$\pi = f_{geo} \circ \pi_0 : \Psi \to \partial\Gamma,$$

where $f_{geo} : \partial T_{geo} \to \partial\Gamma$ is the map associated to the tree T_{geo} which has been defined in Chapter 5. Let $R(\pi)$ be the kernel of π, that is,

$$R(\pi) = \{(\psi_1, \psi_2) \in \Psi \times \Psi \mid \pi(\psi_1) = \pi(\psi_2)\}.$$

We shall show that $R(\pi)$ is a semi-Markovian subset of

$$\Sigma(\mathbf{N}, S) \times \Sigma(\mathbf{N}, S) = \Sigma(\mathbf{N}, S \times S).$$

For that, the following proposition will be useful. The statement and the proof of this proposition are of the same type as those of Proposition 2.2 above.

Proposition 3.1. — *Let y_1 and y_2 be two distinct vertices of T_{geo} such that if they are considered as elements of Γ, they satisfy $\mid y_1 \mid = \mid y_2 \mid$ and $\mid y_1 - y_2 \mid \leq 4\delta$. Then there exists an integer N_2 (which depends only on Γ and on G) such that for every $N \geq N_2$, the vertices y_1 and y_2 are not $(N - 4\delta)$-equivalent.*

PROOF. Let s_0 be a simplicial path of length $\leq 4\delta$ in the Cayley graph K joining the vertices y_1 and y_2. Let γ be the word in the alphabet $G \cup G^{-1}$ which corresponds to the path s_0 when we move along this path, going from y_1 to y_2. Let us consider, starting from the vertex y_1, the simplicial path s labelled by the word $\gamma\gamma\gamma...\gamma$ (γ repeated L times, for a certain integer $L > 0$). If s_0 has length l_0, the path s has length $l_0 L$. As in the proof of Proposition 2.2, if y_1 and y_2 have the same N-type, and if the length of s is less than N, then (the image of) this path in K is contained between the two spheres which arecentered at the identity element of the group and such that the radius of the largest sphere (resp. the smallest one) is equal to the distance of (the image of) s_0 to the identity element (resp. of the largest distance between a point on s_0 and the identity). On the other hand, s is a (λ, k)-quasi-geodesic, and its length is bounded above, as in the proof of Proposition 2.2, by a constant which depends only on λ and on k. Taking N_2 greater than all the constants associated to the different values of λ and k, for all the different possible choices of γ (which are finite in number, because the length of γ is bounded), we are assured that the N_1-types of y_1 and y_2 are different. ∎

As we have already said above, we shall take, in the definition of the set S, $N \geq N_2$.

Definition 3.2. — Let (s_1, s_1') and (s_2, s_2') be two elements of $S \times S$. We say that (s_2, s_2') *follows* (s_1, s_1') in T_{geo} if these two ordered pairs can be represented respectively by two ordered pairs of vertices (y, y') and $(x, x') \in T_{geo} \times T_{geo}$ such that $\mid x \mid = \mid x' \mid$, y follows x and y' follows x' (in the sense defined in §2) (*cf.* Figure 3).

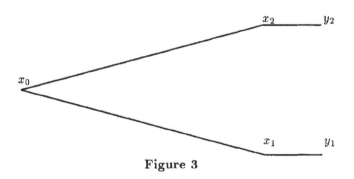

Figure 3

Let now $\Psi_0' \subset \Psi_0 \times \Psi_0$ be the set of sequences of the form $\big(g(n), g'(n)\big)_{n \in \mathbf{N}}$

such that $(g(n))_{n \in \mathbb{N}}$ and $(g'(n))_{n \in \mathbb{N}}$ are discrete geodesic rays in T_{geo} (in the sense defined in §2) which define the same point in ∂X, that is, which satisfy the following condition:

$$f_{geo}(lim_{n \to \infty} g(n)) = f_{geo}(lim_{n \to \infty} g'(n)).$$

From the map $P : \Psi_0 \to \Sigma$ defined in §2, we construct a product map $P' : \Psi'_0 \to \Sigma \times \Sigma$ by formula:

$$P'(g(n), g'(n)) = (P(g(n)), P(g'(n))).$$

Proposition 3.3. — *The map $P' : \Psi'_0 \to \Sigma \times \Sigma$ is injective.*

PROOF. This is a consequence of the fact that the map $P : \Psi'_0 \to \Sigma$ is itself injective. ∎

Let $\Phi' \subset \Sigma \times \Sigma$ be the set of sequences of ordered pairs $(s_n, s'_n)_{n \in \mathbb{N}}$ which satisfy the following condition:

For every $n \geq 0$, any two ordered pairs of consecutive elements of the sequence (s_n, s'_n) and (s_{n+1}, s'_{n+1}) can be represented respectively by two ordered pairs of vertices in the Cayley graph K, (x, x') and (y, y'), such that y follows x and y' follows x' (and therefore (s_{n+1}, s'_{n+1}) follows (s_n, s'_n) in the sense of Definition 3.2), these four vertices of K satisfying furthermore

$$| x - x' | \leq 4\delta$$

and

$$| y - y' | \leq 4\delta.$$

Proposition 3.4. — $\Phi' \subset \Sigma \times \Sigma$ *is a subshift of finite type.*

PROOF. It is clear, from the definition of Φ', that it is a subshift of order two. ∎

Let now $\Psi' = P'(\Psi_0) \subset \Sigma \times \Sigma$ be the image of the map P'. The definition of Ψ'_0 shows immediately that Ψ' is the kernel $R(\pi)$ of π.

Proposition. — Ψ' *is a semi-Markovian subset of $\Sigma \times \Sigma$.*

PROOF. Let us show that Ψ' is equal to the intersection of Φ' with the cylinder

$$C' = \{(s_n, s'_n)_{n \in \mathbb{N}} \subset \Sigma \times \Sigma \text{ with } s_0 = s'_0 = \text{ the } N\text{-type of the basepoint } x_0\}.$$

Let $(s_n, s'_n)_{n \in \mathbb{N}}$ be an element of $\Sigma \times \Sigma$ which is in the kernel of π. The two sequences $(s_n)_{n \in \mathbb{N}}$ and $(s'_n)_{n \in \mathbb{N}}$ are then the images by the map P of two discrete geodesic rays $g(n)$ and $g'(n)$ in the tree T_{geo} which, by definition, converge to the same point of ∂T_{geo}. These two rays start at the basepoint x_0, and we therefore have $s_0 = s'_0 = $ the N-type of x_0. Hence $(s_n, s'_n) \in C'$. On the other hand, by the inequality (3.4.1) of Chapter 1, each of the two geodesic rays stays uniformly at a distance $\leq 4\delta$ from the other one, and we therefore have $(s_n, s'_n)_{n \in \mathbb{N}} \in \Phi'$.

Conversely, let us take an arbitrary element $(s_n, s_n')_{n \in \mathbf{N}}$ in $\Phi' \cap C'$, and let us show that it is in the kernel of π.

For that, we shall construct two discrete geodesic rays $g(n)$ and $g'(n)$ such that $\mid g(n) - g'(n) \mid \leq 4\delta$ for every $n \in \mathbf{N}$, and such that $(s_n, s_n')_{n \in \mathbf{N}} = P'(g(n), g'(n))_{n \in \mathbf{N}}$. The construction of $g(n)$ and $g'(n)$ is made by induction. We begin by defining $g(0) = g'(0) = x_0$, which is possible since $s_0 = s_0' = $ the N-type of x_0.

For every $n \geq 0$, we define the pair $(g(n+1), g'(n+1))$ out of the pair $(g(n), g'(n))$ in the following manner. From the definition of Φ', we can find in the Cayley graph K two ordered pairs of vertices, (x, x') and (y, y') satisfying the following four properties:

- $\mid x \mid = \mid x' \mid$,

- The N-types of these ordered pairs of vertices are respectively the pairs (s_n, s_n') and (s_{n+1}, s_{n+1}').

- y follows x and y' follows x'.

- $\mid x - x' \mid \leq 4\delta$, and $\mid y - y' \mid \leq 4\delta$.

We have, as an induction hypothesis, $\mid g(n) - g'(n) \mid \leq 4\delta$.

Let us consider the left-translation τ of the group which sends x on $g(n)$, and let x'' be the image of x' by this translation.

We have $\mid x'' \mid = \mid g(n) \mid = \mid g'(n) \mid$. Indeed, this is a consequence of the fact that x' is in the ball of radius N centered at x (because we have supposed $N \geq 4\delta$), and that the points x and $g(n)$ have the same N-type.

Let us note now that the translation τ, which preserves the N-type at x, preserves the $(N - 4\delta)$-type at x'. Thus, the points x'' and $g'(n)$ have the same $(N - 4\delta)$-type. By Proposition 3.1, we deduce that $\tau(x') = x'' = g'(n)$.

As τ preserves the $(N - 4\delta)$-type at the points x and x', and as we have taken N large enough (so that τ preserves the isometry types of the pointed trees $T_{geo,x}$ and $T_{geo,x'}$), we conclude that $\tau(y)$ follows x and that $\tau(y')$ follows x'. Applying now Proposition 2.4, we see that $\tau(y)$ has the same N-type than y and that $\tau(y')$ has the same N-type than y'. We then take $g(n + 1) = \tau(y)$ and $g'(n + 1) = \tau(y')$, which completes the induction argument, and the proof of Proposition 3.5. ∎

We can now state the

Theorem 3.6. — *The map $\pi : \Psi \to \partial \Gamma$ is a semi-Markovian presentation of $\partial \Gamma$.*

PROOF. This is a consequence of Propositions 2.7 and 3.5. ∎

Corollary 3.7. — *The boundary of a torsion-free hyperbolic group is a semi-Markovian space.* ∎

Corollary 3.8. — *Let X be a hyperbolic space which is geodesic and proper. Suppose there exists a torsion-free group which acts isometrically and properly discontinuously on X, with X/Γ compact. Then the boundary of X is a semi-Markovian space.*

PROOF. We have $\partial X = \partial \Gamma$, by Proposition 6.5 of Chapter 1. ∎

Corollary 3.9. — *Let Γ be a hyperbolic group which has a finite-index subgroup $\Gamma' \subset \Gamma$ which is torsion-free. Then, the boundary of Γ is a semi-Markovian space.*

PROOF. We know that Γ' is also hyperbolic and that $\partial\Gamma = \partial\Gamma'$ (*cf.* Chapter 1, §2). ∎

Corollary 3.10. — *Let k be a field of zero characteristic, n a natural integer and Γ a hyperbolic group isomorphic to a subgroup of $GL_n(k)$. Then, the boundary of Γ is a semi-Markovian space.*

PROOF. Selberg's lemma (see [Sel] or [Cas] for a proof) asserts that any finitely generated subgroup of $GL_n(k)$ contains a finite-index subgroup which is torsion-free. ∎

§4 – The boundary of $\mathbf{T_{part}}$ as a semi-Markovian subset

To see the space ∂T_{part} as a semi-Markovian subset of a certain one-sided Bernoulli shift, we shall follow the same outline than in §2, concerning the space ∂T_{geo}. We can adapt, step by step, the arguments that we used for the vertices of T_{geo} into arguments for the vertices of T_{part}.

Proposition 4.1. — *Let $N_1 = 60\delta + 2$, let x be an arbitrary vertex of the tree T_{part} and let y_1 and y_2 be two distinct vertices of this tree which are situated beyond x and at distance 1 from x. Then, there exists an integer $N_0 \geq N_1$ (which deepends only on the group Γ equipped with its given set of generators) such that for every $N \geq N_0$, the vertices y_1 and y_2 are not $(N - N_1)$-equivalent.*

PROOF. Let us identify the vertices y_1 and y_2 with their images as finite subsets of Γ. If $card(y_1) \neq card(y_2)$, these two vertices are not N-equivalent, for any $N \geq 0$. If $card(y_1) = card(y_2)$, we can find two elements z_1 and z_2, with $z_1 \in y_1 - y_2$ and $z_2 \in y_2 - y_1$. The points z_1 and z_2 are situated on the same sphere centered at the identity element. As the diameters of the three sets x, y_1 and y_2 of Γ are bounded above by 20δ, we see that the distance $\mid z_1 - z_2 \mid$ is bounded above by $3 \times 20\delta + 2 = 60\delta + 2$. In the same manner as for the proof of Proposition 3.1, we can see now the existence of the integer N_0 which we are looking for. ∎

We then proceed in the same way as we did in §2, concerning the tree T_{geo}: we define the set of symbols S as the set of N-equivalence classes of vertices of the tree T_{part}, and if s_1 and s_2 are two elements of S, we say that s_2 *follows* s_1 in T_{part} if these two elements can be represented by vertices x' and y' of that tree, with $y' \in T_{part,x'}$ and $\mid x' - y' \mid = 1$.

We have then the following proposition, which is analogous to Proposition 2.4.

Proposition 4.2. — *Let s_1 and s_2 be two elements of S such that s_2 follows s_1 in T_{part}. If $x \in s_1$ is an arbitrary vertex of T_{part} in the class s_1, then there exists a vertex y which follows x and which is in the class s_2. In fact, if x' and y' are as above and if we identify the four points x, y, x' and y' with their image sets in Γ, then the* *image of y' by the left-translation of Γ which sends x' on x is a subset of Γ which defines the vertex y of T_{part}.*

PROOF. (sketch). The proof is of the same type as that of Proposition 2.4. We use Proposition 1.9 which insures that the pointed trees $T_{part,x}$ and $T_{part,x'}$ are N-equivalent. ∎

The rest of the construction is exactly the same as for the tree T_{geo}:

We define the set Ψ_0 of discrete geodesic rays in T_{part} which start at the basepoint, and a map

$$P : \Psi_0 \to \Sigma = \Sigma(\mathbf{N}, S)$$

which, thanks to Proposition 4.1, will be injective.

The image $\Psi \subset \Sigma(\mathbf{N}, S)$ of Ψ_0 by P is a semi-Markovian subset of $\Sigma(\mathbf{N}, S)$.

There is a natural bijection $D : \Psi_0 \to \partial T_{part}$, and if we call π_0 the composed map:

$$\pi_0 = D \circ P_0^{-1} : \Psi \to \partial T_{part},$$

we have the following

Theorem 4.3. — *The map π_0 is a homeomorphism between the semi-Markovian subset $\Psi \subset \Sigma(\mathbf{N}, S)$ and ∂T_{part}. In particular, π_0 is a semi-Markovian presentation of ∂T_{part}.* ∎

§5 – A finite-to-one semi-Markovian presentation of $\partial\Gamma$

From the semi-Markovian presentation of ∂T_{part} decribed in §4 above, we can deduce a semi-Markovian presentation of $\partial\Gamma$ following the same type of reasoning as for the presenattion give in §3. The set of symbols S will be here the set of N-equivalence classes of vertices of T_{part}, with N greater than the integer N_0 of Proposition 4.1 and the integer N_1 of Proposition 5.1 below which will be useful for the construction of the presentation, and which is the analogue of Proposition 3.1.

Proposition 5.1. — *Let y_1 and y_2 be two distinct vertices of T_{part}, such that $\mid y_1 \mid = \mid y_2 \mid$ and such that if we consider these vertices as finite subsets of Γ, we have $dist(y_1, y_2) \leq 4\delta$. Then there exists an integer N_1 such that for all $N \geq N_1$, the vertices y_1 and y_2 are not $(N - 4\delta)$-equivalent.*

PROOF. The proof is of the same style as that of Propositions 4.1 and 3.1 (in fact, we can see this proposition as a corollary of Proposition 3.1). ∎

We then proceed in the construction of the semi-Markovian presentation of $\partial\Gamma$. Following the same notations as in §3, we define the set $\Phi' \subset \Sigma \times \Sigma$ as the set of sequences of ordered pairs $(s_i, s_i')_{i \in \mathbf{N}}$ satisfying the following condition:

$\forall n \geq 0$, two consecutive ordered pairs (s_n, s'_n) and (s_{n+1}, s'_{n+1}) can be represented by ordered pairs of vertices of the tree T_{part}, (x, x') and (y, y') such that $\mid x \mid = \mid x' \mid$, y following x and y' following x', and such that if these four vertices are considered as finite subsets of Γ, we have $dist(x, x') \leq 4\delta$ and $dist(y, y') \leq 4\delta$.

The other definitions and the rest of the construction remain inchanged with respect to §3, and we thus obtain the following

Theorem 5.2. — *The map $\pi : \Psi \to \partial\Gamma$ which is defined in this way is a semi-markovian presentation of $\partial\Gamma$ which is finite-to-one, that is, there exists an integer M such that, for every $\xi \in \partial\Gamma$, $card(\pi^{-1}(\xi)) \leq M$.*

Notes and comments on Chapter 7

Proposition 8.5.K of [Gro 3] affirms that the boundary of every hyperbolic group (without the hypothesis on torsion) is semi-Markovian. Gromov gives indications for the proof in sections 8.5.I and 8.5.J of [Gro 3].

Bibliography for Chapter 7

[Can] J. Cannon, "The combinatorial structure of co-compact discrete hyperbolic groups", *Geometriae Dedicata* **16**, (1984), pp. 123-148.

[Cas] J. W. S. Cassels, "An embedding theorem for fields", *Bull. Australian Math. Soc.* **14**, (1976), pp. 193-198 and 479-480.

[CDP] M. Coornaert, T. Delzant, A. Papadopoulos, "Geométrie et théorie des groupes: Les Groupes hyperboliquers de Gromov", Lecture Notes in Mathematics, vol. 1441, Springer Verlag, 1990.

[Gro 1] M. Gromov, "Hyperbolic manifolds, groups and actions", *Ann. of Math. Studies* **97**, Princeton university Press (1982), pp. 183-215.

[Gro 3] ——, "Hyperbolic groups", *in* Essays in Group Theory, MSRI publ. **8**, Springer, 1987, pp. 75-263.

[Sel] A. Selberg, "On discontinuous groups in higher dimensional spaces", *in "Contributions to Function Theory"*, Bombay 1960, pp. 147-164.

Index

Bernoulli shift 20
  one-sided — 31
  two-sided — 31
bounded geometry (simplicial graph with ——) 60
boundary
  — of a hyperbolic group 15
  — of a hyperbolic space 9
Busemann function 46

canonical metric on a graph 15
Cantor set 20
Cayley graph 15
cocompact action 15
cocycle φ 44
convergent quasi-geodesic field 74
curve
  Menger — 113
  Sierpinski — 112
cylinder 26

diagonal 22
dimension (topological ——) 104
dynamical system 22

elliptic isometry 16
expansive system 22
expansivity constant 22

finite presentation 29
finite type
  subshift of —— 28

system of ——	29
function	
(ζ- ——)	35
(Busemann ——)	46
geodesic	
— polygon	7
— ray	6
— segment	6
— space	6
gradient line	50
graph	
— with labelled vertices	36
simplicial —— with bounded geometry	60
Cayley ——	15
transition ——	34
Gromov product	6
group	
— acting cocompactly	15
— acting properly discontinuously	15
hyperbolic ——	8
Hausdorff distance	14
hyperbolic	
— group	8
— isometry	16
— metric space	6
internal points of a triangle	7
isometry	
elliptic ——	16
hyperbolic ——	16
parabolic ——	16
line (gradient ——)	50
Lipschitz map	16
Markovian subshift	28
Menger	
— curve	113
— sponge	113
matrix (transition ——)	32
metric	
canonical —— on a simplicial complex	15
word ——	8
visual —— on the boundary	12

metric space
  geodesic ——   6
  hyperbolic ——   6
  proper ——   10

narrow (δ-) geodesic polygon   7
N-type   119
N-equivalence   119

parabolic isometry   16
perfect set   20
point at infinity of a cocycle φ 50
pointed subtree   120
polygon (geodesic ——)   6
polyhedron $P_d(X)$ 17
presentation (finite ——) of a dynamical system   28
presentation (semi-Markovian ——) of a compact set 108
primitive of a cocycle φ 45
product (Gromov ——)   6
projection   96
projective sequence of sets   92
properly discontinuous action   15
proper metric space   10

quasi-geodesic   13
quasi-isometry   13
quotient system   24

real tree   6

segment
  geodesic ——   6
  topological ——   6
semigroup   19
semi-Markovian
  —— presentation   108
  —— space   108
  —— subset   108
sequence
  convergent —— at infinity   9
  projective —— of connected sets   92
set of symbols   20
Sierpinski
  —— carpet   112
  —— set   112
  —— sponge   112

sofic
  — subshift   36
  — system   37
subshift
  — of finite type   28
  — of order n 31
  Markovian ——   28
  sofic ——   36
symbols (set of ——)   20
system
  — of finite type   29
  dynamical ——   22
  expansive ——   22
  finitely presented ——   29
  quotient ——   24
  sofic ——   36

theorem
  — of approximation by trees   13
  — of stability of quasi-geodesics   14
  Manning's rationality ——   36
thin (δ- —— triangle)   7
totally disconnected space   20
topological dimension 0   25
topological dimension n 104
topologically conjugate systems   25
transition
  — graph   34
  — matrix   33
transitive action   21

visual metric   12

Printing: Druckhaus Beltz, Hemsbach
Binding: Buchbinderei Schäffer, Grünstadt